# Practical Guide to Inspection, Testing and Certification of Electrical Installations

## Conforms to IEE Wiring Regulations/BS 7671/Part P of Building Regulations

Second Edition

**Christopher Kitcher**

ELSEVIER

AMSTERDAM • BOSTON • HEIDELBERG • LONDON • NEW YORK • OXFORD
PARIS • SAN DIEGO • SAN FRANCISCO • SINGAPORE • SYDNEY • TOKYO
Newnes is an imprint of Elsevier

Newnes

Newnes is an imprint of Elsevier
Linacre House, Jordan Hill, Oxford OX2 8DP, UK
30 Corporate Drive, Suite 400, Burlington, MA 01803, USA

First edition 2008
Second edition 2009

Notice
No responsibility is assumed by the publisher for any injury and/or damage to persons or property as a matter of products liability, negligence or otherwise, or from any use or operation of any methods, products, instructions or ideas contained in the material herein.

Disclaimer
This book draws on many sources. Some are facts, some hypotheses, some opinions. Most – including many of my own statements – are mixtures. Even 'facts' are unavoidably selective and can rarely be guaranteed. Despite careful checking, neither I nor my colleagues or publishers can accept responsibility for any errors, misinformation or unsuitable advice. This also applies to opinions – particularly on issues affecting health and safety. As any recommendations must balance complex, often opposing, factors, not everyone will reach the same conclusions. In this – as indeed in every issue this book touches on – every reader must make up her or his mind, for which they alone must be responsible.

I offer the best advice I am capable of, but every circumstance is different. Anyone who acts on this advice must make their own evaluation, and adapt it to their particular circumstances.

**British Library Cataloguing in Publication Data**
A catalogue record for this book is available from the British Library

**Library of Congress Cataloging-in-Publication Data**
A catalog record for this book is available from the Library of Congress

ISBN: 978-1-85617-607-1

For information on all Newnes publications
visit our website at www.Elsevier.com

Printed and bound in Great Britain

09 10 11 12 13   10 9 8 7 6 5 4 3 2 1

# Contents

# Preface

Part P of the Building Regulations came into effect on the 1st of January 2005. This part of the Building Regulations requires that all electrical work carried out in domestic installations now has to be certificated. For work places, the Electricity at Work Regulations 1989 require that they provide a safe and well maintained electrical system. Certification and well kept records are a perfect way to confirm that every effort has been made to ensure that the system is, and remains safe.

I have written this book to assist electricians of all levels in carrying out the inspecting, testing and certification of all types of electrical installations. It will also be invaluable to **City and Guilds 2330** level 2 and 3 students, electricians studying for **City and Guilds 2391-10, 2391-20, 2392-20** and all tradesmen who are required to comply with **Building Regulation part P**.

Step by step guidance and advice is given on how to carry out a detailed visual inspection during initial verifications, periodic inspections and certification.

Text and colour photographs of real, not simulated installations, are used to show the correct test instruments. Step by step instructions for how to carry out each test safely using different types of instruments are given along with an explanation as to why the tests are required.

In some photographs safety signs have been omitted for clarity. All test leads are to GS 38 where required. Some comments within this book are my own view and have been included to add a common sense approach to inspecting and testing.

Interpretation of test results is a vital part of the testing process. The correct selection and use of tables from BS 7671 and the On Site Guide are shown clearly. Any calculations required for correct interpretation or for the passing of exams are set out very simply. An in depth knowledge of maths is not required.

I have included questions and example scenarios, along with answers and completed documentation. These will assist electricians at all levels whether they need to pass an exam or complete the certification.

During my time lecturing I have been asked many questions by students who have become frustrated by being unable to reference definitive answers. Within this book I have tried to explain clearly and simply many of these difficult questions.

*Chris Kitcher*

# Foreword

Christopher Kitcher is a very experienced electrician and teacher. I have worked with him for the last 11 years in both the college environment and 'on site'. We are both examiners and also work for the City of Guilds on the 2391 qualification. Christopher now mainly works in education as well as writing books; this is his second book in the electrical installation sector, his first was the successful and very practical update to the WATKINS series.

Christopher's track-record speaks for it self; his work was instrumental in his college gaining the status of Centre of Vocational Excellence this is an accolade for training providers who gain a grade 1 or 2 during the joint OFSTED/ALI inspection.

*Practical* is again the keyword, as the book takes the reader from why to how in very clear steps. There are not only clear explanations, but photographs to guide the reader.

Whether an experienced electrician testing large industrial or commercial installations or a domestic installer altering or adding to installations, Chris has got it covered for you, generally using a language that installers use and not just technological terms that most installers find it difficult to understand. After all, we are electricians and not English language teachers, aren't we?

This new and exciting practical guide will support candidates who are looking to take exams such as the City and Guilds testing and inspection 2391 or 2, the EAL VRQ level 2 Domestic installer and the NIC Part P course.

*Gerry Papworth B.A. (Hons), MIET, Eng. Tech, LCGI*
Managing Director of Steve Willis Training (Portsmouth) Ltd
Chief examiner for City & Guilds 2391

# Acknowledgements

Writing this book has been a challenge which I have thoroughly enjoyed. It has been made a pleasurable experience because of the encouragement and generosity of many people and organisations. Particular thanks go to:

For the use of test equipment:
Phil Smith of Kewtech
Peter Halloway of Megger UK.

All at NIC certification for their help and assistance in allowing the reproduction of certificates and reports.

My granddaughter Heather Bates whose computer skills and patience saved the day on many occasions!

My colleagues at Central Sussex College for their valuable input, support and expertise:
Dave Chewter
Jason Hart
Lee Ashby
Andy Hay-Ellis
Jon Knight
Simon Nobbs

Brian Robinson for taking the photographs.

Finally, a special thanks is due to Central Sussex College for allowing me to use their facilities and equipment.

*Chris Kitcher*

# Inspection and testing of electrical installations

## WHY INSPECT AND TEST?

The Electricity at Work Regulations 1989 is a statutory document; it is a legal requirement that statutory regulations are complied with. Not to comply is a criminal offence and could result in a heavy fine and even imprisonment in extreme cases.

These regulations are required to ensure that places of work provide a safe, well-maintained electrical system. A simple way to provide this is to ensure that newly installed circuits and existing installations are tested on a regular basis. Electrical test certificates are used to record what has been done and confirm that the installation meets the required standard.

The British standard for electrical installations is BS 7671, the requirement for electrical installations. Within this standard, **Regulation 610.1** states that "every installation shall, during erection and on completion before being put into service be inspected and tested to verify, so far as reasonably practicable, that the requirements of the regulations have been met".

**Regulation 621.1** states that "where required, periodic inspection and testing of every electrical installation shall be carried out in accordance with regulations 621.2 to 621.5 in order to determine as far as is reasonably practicable, whether the installation is in a satisfactory condition for continued service".

**Document P** of the Building Regulations 2000 for Electrical Safety came into effect on 1 January 2005 and was amended in April 2006.

The purpose of this document is to ensure electrical safety in domestic electrical installations.

## Section 1. Design, Installation, Inspection and Testing

This section of Part P is broken down into sub-sections.

### General

This states that electrical work must comply with the Electricity at Work Regulations 1989 and that any installation or alteration to the main supply must be agreed with the electricity distributor.

### Design and installation

This tells us that the work should comply with BS 7671 electrical wiring regulations.

### Protection against flooding

The distributor must install the supply cut out in a safe place and take into account the risk of flooding. Compliance with The Electrical Safety, Quality and Continuity Regulations 2002 is required.

### Accessibility

Part M of the building regulations must be complied with.

### Inspection and testing before taking into service

This area is covered in detail throughout this book, it reminds us that the installation must be inspected and tested to verify that it is safe to put into service.

### BS 7671 Installation certificates

This tells us that compliance with Part P can be demonstrated by the issue of a correct Electrical Installation Certificate and also what the certificate should cover. This is addressed later in this book.

### Building regulation compliance certificates or notices for notifiable work

This tells us that the completion certificates issued by the local authorities, etc. are not the same as the certificates that comply with BS 7671. The completion certificates do not only cover Part P, but also shows compliance with all building regulations associated with the work which has been carried out.

### Certification of notifiable work

This is covered in detail throughout this book.

### Inspection and testing of non-notifiable work

This tells us that, even if the work is non-notifiable, it must be carried out to comply with BS 7671 and that certificates should be completed for the work.

### Provision of information

Information should be provided for the installation to assist with the correct operation and maintenance. This information would comprise of certification, labels, instruction and plans.

## Section 2. Extensions, Material alterations and material changes of use

This section is covered throughout this book, it basically tells us that certification is required, and that before any additions or alterations are made to an installation, an assessment of the existing installation should be made, to ensure that it is safe to add to.

## Section 3. Information about other legislation

This covers the Electricity at Work Regulations 1989; Electrical Safety, Quality and Continuity Regulations 2002; functionality requirements.

The construction design and management regulations also state that adequate electrical inspection and tests are carried out on all new installations, those with electrical design information shall form a user's manual, which can be used to provide an up-to-date working record of the installation.

With the introduction of the 'Home Information Pack' (HIP) selling a property will eventually become very difficult if not impossible unless all of the relevant documentation is in place, this of course will include certification of electrical systems. Whilst, at the time of writing, this certification is not a requirement of the HIP, it is almost certain to become so in the future. Mortgage lenders and insurance companies are frequently asking for certification as part of the house buying/selling process. Owners of industrial and commercial properties could find that insurance is difficult to obtain, while most licensing bodies and local authorities are asking for electrical certification within their guidelines.

All of these regulations are under the umbrella of the Health and Safety at Work Act 1974. This clearly puts the legal responsibly of health and safety on all persons.

### COMPLIANCE WITH BUILDING REGULATIONS PART P

Compliance with building regulations is a legal requirement and electrical work carried out in the domestic sector is now included in the building regulations; it is a criminal offence not to comply with the building regulations.

At the time of writing, there is no legal requirement to notify any work carried out in commercial or industrial buildings, although it should still be certificated for safety and record-keeping purposes.

Document P requires that most electrical work carried out in domestic premises is notified to the local authority building control. There are a few exceptions but the work must comply with BS 7671 Wiring Regulations. The exceptions are as follows:

*Minor works carried out in areas that are not classed as special locations and therefore do not need notifying but would still need certifying*
- Addition of socket outlets and fused spurs to an existing radial or ring circuit.
- Addition of a lighting point to an existing circuit.
- Installing or upgrading main or supplementary bonding.

*Minor works carried out in the Special Locations as listed below – or in Kitchens* (BS 7671 does not recognize a Kitchen as a special location. Document P does)
    Kitchens
    Locations containing bath tubs or shower basins
    Hot air saunas
    Electric floor or ceiling heating
    Garden lighting (*if fixed to a dwelling wall it is not deemed to come into the Special Location category*)
    Solar photovoltaic power supply systems

The work which could be carried out in these locations without notification but should still be certificated would be:

- Replacement of a single circuit which has been damaged
    *Providing that the circuit follows the same route*
    *The cable used has the same current carrying capacity as the cable being replaced*
    *Circuit protective measures are not affected.*
- Replacing accessories such as socket outlets, switches and ceiling roses.
- Re-fixing or replacing of enclosures and components.

All other work carried out in any areas of a domestic installation must be certificated and notified to the local authority building control, this can be carried out by various methods.

## EARTHING AND BONDING TO COMPLY WITH PART P

If a Minor Electrical Installation Works Certificate is necessary, there is no requirement to upgrade the existing earthing and bonding arrangements within an installation. Where the earthing and bonding do not comply with the latest edition of BS 7671, it should be recorded on the Minor Electrical Installation Works Certificate.

If an Electrical Installation Certificate is required, then the earthing arrangements must be upgraded to comply with the current edition of BS 7671.

Where the work is in the bathroom, or any areas that require supplementary bonding, then this must also be brought up to the current standard.

There is no requirement to upgrade supplementary bonding in an area where work is not to be carried out. There is also no requirement under Part P to certificate the upgrading of earthing and bonding to an installation.

## REGISTERED DOMESTIC INSTALLER

To become a registered domestic installer, it is necessary to become a member of one of the certification bodies which operate a domestic installer's scheme. This would require the person carrying out the work to prove competence in the type of work which is being carried out, and the ability to inspect, test and certificate the work which he/she has carried out. Competence is usually assessed by a site visit from an inspector employed by the chosen scheme provider.

There are two types of registration: (1) a person who needs to be able to carry out all types of electrical installation work in dwellings will need to register with an organization which runs a full scope scheme; (2) a person who needs to carry out electrical work associated with their main trade will need to register with an organization which runs a limited scope scheme. This scheme will enable a person to carry out electrical work which is related to the other work which is being carried out. An example of this would be where a person is a Kitchen fitter and needs to carry out electrical work which is required in the Kitchen. The installer would not be allowed to carry out electrical work in other parts of the dwelling unless that person was a member of a full scope scheme.

If the electrician is registered as a domestic installer, he or she must complete the correct certification and notify the scheme provider, who they are registered with, of the work which has been carried out. This must be done within 30 days. The scheme provider will both notify the local authority and the customer of the correct certification being given. An annual fee is usually required by the scheme provider, while a small fee is also payable for each job registered.

### UNREGISTERED COMPETENT PERSON

If the work is carried out by a non-registered competent person who is capable of completing the correct certification, the local authority will need to be contacted before commencement of work, and the work will be carried out under a building notice. This will involve a fee being paid to the local authority and a visit or visits being made by a building inspector to inspect the work being carried out to ensure that it meets the required standard (*the cost of this will usually be far higher than that charged per notification by a scheme provider to a registered installer*). On satisfactory completion, and after the issue of the correct certification by the competent person, the building inspector will issue a completion certificate. The issue of a completion certificate by the local authority does not remove the responsibility for the work including guarantees from the non-registered competent person; the required certification must still be completed by the person who carried out or who is responsible for the work.

### DIY INSTALLER

In cases where the work is carried out by a person who could not be deemed qualified (i.e. a DIY enthusiast), building control must be informed prior to work commencing, and on completion of the work to the building control officer's satisfaction, an inspection and test certificate must be issued. As a DIY installer would be unlikely to have the knowledge, experience or correct test equipment required to carry out the inspection, tests or completion of the certification, the services of a competent person would be required. The qualified person would in effect take responsibility for the new/altered work. For that reason, the qualified person would need to see the work at various stages of the installation to verify that the work and materials used comply with the required standards of the BS 7671 wiring regulations.

### SUMMARY

Currently, there is no requirement for any person carrying out electrical work in a domestic environment to be qualified in any way. The condition is that they must be competent; in other words, they must be in possession of the appropriate technical knowledge or experience to enable them to carry out the work safely.

There are Part P courses being provided by many training bodies, although it is not a requirement that you attend one of these courses or any other course which is being offered. However, it is impossible to become an electrician in 5 days.

The buildings control authorities must be informed of any electrical work that is to be carried on a domestic electrical installation other than very minor work, although even this work must be certificated.

Building control can be informed (*before commencing work*) by the use of a building notice, and this will involve a fee.

If your work involves a lot of domestic electrical work, then by far the best route would be to join one of the certification bodies. This would allow you to self-certificate your own work. When you join one of these organizations, you must be able to show that your work is up to a satisfactory standard and that you can complete the correct paperwork (*test certificates*). Whichever organization you choose to join, they will give you the correct advice on which training you require. A qualification is fine, but being able to carry out electrical work safely is far better.

# 2

# Types of certification required for inspecting and testing of installations

## CERTIFICATION REQUIRED FOR DOMESTIC INSTALLATIONS (PART P)

The certification requirements for compliance with Part P are similar to the conditions for any other electrical installation.

It is a legal requirement to complete a Minor Electrical Installation Works Certificate (commonly called a 'Minor Works Certificate' or an 'Electrical Installation Certificate' for any electrical work being carried out on a domestic installation).

### Minor Electrical Installation Works Certificate

This is a single document that must be issued when an alteration or addition is made to an existing circuit. A typical alteration that this certificate might be used for is the addition of a lighting point or socket outlet to an existing circuit. This certificate would be used for any installation regardless of whether it is domestic or not.

### Part P, Domestic Electrical Installation Certificate

An Electrical Installation Certificate is required for:

- A new installation.
- When new circuits are installed.
- When a single circuit is installed.
- The changing of a consumer's unit.
- When a circuit is altered and the alteration requires the changing of the protective device.

This document is usually made up of three parts: (1) the Electrical Installation Certificate; (2) the Schedule of Inspections, and (3) the Schedule of Test Results (see Chapter 5). The format of these documents will differ slightly depending on who they are supplied by, but the content and legal requirement is the same.

### Periodic inspection, testing and reporting

There is no requirement in Part P for periodic inspection, testing and reporting. However, if the replacement of a consumer's unit has been carried out, then the circuits which are reconnected should be inspected and tested to ensure that they are safe. This will, of course, require documentation: a Periodic Inspection Report, Schedule of Test Results and a Schedule of Inspection.

It is not a requirement of Part P that specific Part P certificates are used but you will find that many clients/customers prefer them.

The certificates produced by the IET (previously known as the IEE) are sufficient to comply with Part P and can be downloaded from www.theiet.org as described in the general certification section.

Some documents contain a **schedule of items tested**, which can also be found on the IET website. Although it is not a requirement that this document is completed, it is often useful as a checklist.

### CERTIFICATION REQUIRED FOR THE INSPECTING AND TESTING OF INSTALLATIONS OTHER THAN DOMESTIC

(Further explanation is provided for these documents later in the book)

All of these certificates are readily available from many sources. The basic forms can be downloaded from the IET website which is www.theiet.org. Once on the site click on *publication*; next on *BS7671 Wiring Regulations*, then on *Forms for Electrical Contractors*, and this will take you to all of the forms. If you scroll down the page a package is available that will allow you to fill in the forms before printing them.

The NICEIC have forms which can be purchased by non-members and most instrument manufacturers produce their own forms, which are also available from electrical wholesalers.

## Minor Electrical Installation Works Certificate

This is a single document which should be issued if any alteration or addition is made to an existing circuit such as an additional lighting point or spurred socket outlet.

## Electrical Installation Certificate

This certificate must be issued for a completely new installation or new circuit; this would include alterations to a circuit which would result in the changing of a protective device or the renewal of a distribution board.

The Electrical Installation Certificate must be accompanied by a Schedule of Test Results and a Schedule of Inspection. Without these two documents, the Electrical Installation Certificate is not valid. This certificate must not be issued until the installation complies with BS 7671.

An inspection and test which is carried out on a new installation to prove compliance is called an <u>initial verification</u>.

## Initial verification inspection

The documentation which should be completed is the Electrical Installation Certificate; this must be accompanied by a Schedule of Test Results and a Schedule of Inspection.

The purpose of this inspection is to verify that the installed equipment complies with BS or BS EN standards; that it is correctly selected and erected to comply with BS 7671; and that it is not visibly damaged or defective so as to impair safety (*Regulation 611.2*).

When a new installation has been completed, it must be inspected and tested to ensure that it is safe to use. This process is known as the initial verification (Regulation 610.1). For safety reasons, the inspection process must precede testing.

Regulation 610.1 clearly tells us that the inspecting and testing process must be ongoing from the moment the electrical installation commences. In other words, if you are going to be responsible for completing the required certification, you must visually inspect any parts of the installation which will eventually be covered up.

> You should never certificate any work which you have not seen during installation; once you sign the certificate you will be accepting a level of responsibility for it.

For this reason, by the time the installation is completed and ready for certification, a great deal of the installation will have already been visually inspected.

As an initial verification is ongoing from the commencement of the installation, much of the required inspecting and testing will be carried out during the installation, it is important that the whole range of inspection and tests are carried out on all circuits and outlets. Clearly it would not be sensible to complete the installation and then start dismantling it to check things like tight connections, fitting of earth sleeving and identification of conductors, etc.

There are many types of electrical installations and the requirements for them will vary from job to job. Where relevant, the following items should be inspected to ensure that they comply with BS 7671, during erection if possible:

- Have correct erection methods been used?
- Are diagrams and instructions available where required?
- Have warning and danger notices been fitted in the correct place?
- Is there suitable access to consumers' units and equipment?
- Is the equipment suitable for the environment in which it has been fixed?
- Have the correct type and size of protective devices been used?
- Have 30 mA residual current devices been fitted to circuits supplying socket outlets which are likely to be used by ordinary persons (Regulation 411.3.3).
- Where a socket outlet has been installed without RCD protection, is it correctly labelled, has the wiring been installed in the correct zones with an earthed metallic covering where required? (Regulations 411.3.3/522.6.6)
- Have 30 mA residual current devices been fitted to all other circuits where they are required?
- Are the isolators and switches fitted in the correct place?
- Could the installation be damaged by work being carried out on other services or by movement due to expansion of other services?
- Are bands 1 and band 2 circuits separated?
- Have the requirements been met or basic and fault protection?
- Are fire barriers in place where required?
- Are the cables routed in safe zones? If not, are they protected against mechanical damage?
- Are the correct size cables being used, taking into account voltage drop and current carrying requirements?

it is not a requirement that they are, but an RCD should be listed as a recommendation.

- Are earthing clamps to BS 951 standards and correctly labelled?
- If gas, water is bonded using the same conductor, ensure that the conductor is continuous and not cut at the clamp.
- Is the supplementary bonding in place in bathroom? (See Chapters 4–7, and Figure 4 of the On-Site Guide.)
- Is the correct equipment for the correct zones in bath/shower room? (See 701 BS 7671)
- Has the bedroom had a shower installed? If so, are the socket outlets 3 metres from the shower and RCD protected?
- Is there any evidence of mutual detrimental influence; are there any cables fixed to water, gas or any other non-electrical services? (*The cables need to be far enough away to avoid damage if the non-electrical services are worked on.*)
- Are the cables of different voltage bands segregated? Low voltage, separated extra low voltage (SELV), telephone cables or television aerials should not be fixed together (*although they are permitted to cross*).

Whilst these items are being checked, look in any cupboards for sockets or lights. If your customer is uncomfortable with this it is vitally important that you document any areas that cannot be investigated in the extent and limitation section on the Periodic Inspection Report. During this purely visual part of the inspection you will gain some idea of the condition of the installation, and indeed any alterations which have been carried out by a qualified tradesman or by a cowboy/girl.

Clearly, if it is an old installation, an electrical installation certificate must be completed and some of the items listed above will apply. However, if it is a new installation, access to all areas must be secure; if this is not possible then the certificate should not be issued. Again, this list is not exhaustive but will not require removal of any fittings, etc.

Providing that you are happy that the installation is safe to tamper with, a more detailed visual inspection can be carried out and the dreaded but necessary form filling can be started.

Once again begin at the consumer unit. Before you start, this must be isolated. The Electricity at Work Regulations 1989 states that it is an offence to work live. Once you remove a cover you will be working live if you do not isolate it first. Having carried out the safe isolation procedure, remove the cover of the consumer unit.

- Your first impression will be important – has care been taken over the terminations of cables (*neat and not too much exposed conductor*)?

- Are all cables terminated and all connections tight (*no loose ends*)?
- Are there any signs of overheating?
- Is there a mixture of protective devices?
- Are there any rubber cables?
- Are there any damaged cables (*perished or cut*)?
- Have all circuits got Circuit Protective Conductors (CPCs)?
- Are all earthing conductors sleeved?
- On a photocopy of a Schedule of Test Results record circuits, protective devices and cable sizes.
- Look to see if the protective devices seem suitable for the size cables that they are protecting.
- Note any type D or 4 circuit breakers – these will require further investigation.
- Are all barriers in place?
- Have all of the circuit conductors been connected in sequence, with phase, neutral and CPC from circuit number 1 being in terminal number 1 – preferably the highest current nearest the main switch?
- Have any protective devices got multiple conductors in them, are they the correct size (*all the same*)?
- Is there only one set of tails or has another board been connected to the original board by joining at the terminals?

**Having had a detailed look at the consumer unit, and with the installation still isolated**, carry out a more detailed investigation of the rest of the installation.

It may be that you have agreed with your client that only 10% of the installation is to be inspected. This would mean 10% of each circuit. There would be little point in inspecting 10% of the circuits. If the period between inspections was 10 years it could be many years before a circuit was eventually inspected and the exercise would be pointless.

During your preliminary walk around, you will have identified any areas of immediate concern, and these must be addressed as your inspection progresses. There is no reason why you should not start your dead testing at this point, as you progress through your visual inspection.

On radial circuits this would be a good time to carry out CPC continuity, $R_1 + R_2$, insulation resistance and polarity tests as you work your way round. Start at circuit number 1 and work your way through the circuits one at a time.

But first what are you looking for? Let's look at a selection of circuits.

Remember the installation may not comply with the current regulations. If this is the case then a judgement will have to be made by you, as to whether the installation is recorded as satisfactory or unsatisfactory. Clearly an installation which was safe when it was installed does not become unsafe because the regulations have been updated.

Compliance BS 7671:2008 will require that, in most instances, circuits should be RCD protected. It will be some considerable time before this will be found to be the case when carrying out periodic inspections. In most instances this type of non-compliance would be regarded as a 4 in the observations and recommendations section of the PIR.

*Shower circuit*
- Is isolation provided, if so is it within prescribed zones? (*Remember the switch can be anywhere outside of zone 2*)
- Has the correct size cable/protective device been selected?
- Is it bonded?
- Are connections tight?
- Has earth sleeving been fitted?
- Is the shower secure?
- Is there any evidence of water ingress?
- Is the shower in a bedroom? (if it is are the socket outlets RCD protected and 3 meters from the shower?)

Note 1: for any installation carried out after June 2008 any final circuit up to and including 32 ampers must have a maximum disconnection time of 0.4 seconds if connected to a TN system and 0.2 seconds if connected to a TT system.

*Cooker circuit*
- Is the switch within 2 metres of the cooker or hob?
- Has the cooker switch got a socket outlet? If so it requires a 0.4 second disconnection time (note 1).
- Green and yellow sleeving fitted.
- If it has a metal faceplate has it got an earth tail to the flush box?
- Is the cable the correct size for protective device?
- Are there any signs of overheating around the terminations?
- Is the cooker outlet too close to the sink? Building regulations require any outlets installed after January 2005 should be at least 300 mm from the sink.

*Socket outlets*
- Is there correct coordination between protective devices and conductors?
- Green and yellow sleeving fitted.
- Do any metal sockets have an earthing tail back to the socket box?
- Radial circuit not serving too large an area (*see Table 8 A of the On-Site Guide*).
- Secure connections.
- Are cables throughout the circuit the same size?
- Are there any sockets outside? Are they waterproof? Are they 30 mA RCD protected?
- Are there any outlets in the bathroom? If there is, are they SELV? or are they 3 metres from the bath or shower and RCD protected?
- Are there socket outlets within 3 metres of a shower installed in a bedroom? If there is, are they 30 mA RCD protected?
- Will the protective device for the circuit provide 0.4 seconds disconnection time?

*Fused connection units and other outlets*
- As above but could be 5 second disconnection time.
- Does it supply fixed equipment in bathrooms? Are they in the correct zones?
- Do they require RCD protection? This would be all circuits in a bathroom which has had the wiring installed since June 2008
- Functional switching devices shall be suitable for the most onerous duty that they are intended to perform (537.5.2.1)

*Immersion heater circuits*
- Is there correct coordination between the protective device and live conductors?
- Has the CPC been sleeved?
- Is the immersion the only equipment connected to this circuit? (*Any water heater with a capacity of 15 litres or more must have its own circuit.*) *On-Site Guide*, Appendix 8.5. Often you will find that the central heating controls are supplied through the immersion heater circuit.
- Is the immersion heater connected with heat resistant cord?
- The immersion heater switch should be a cord outlet type; not a socket outlet and plug.
- If the supplementary bonding for the bathroom is carried out in the cylinder cupboard, does the supplementary bonding include the immersion heater switch? (It should.)

*Lighting circuits*

- Is there correct coordination between the protective device and the live conductors?
- How many points are there on the circuit? A rating of 100 watts minimum must be allowed for each lighting outlet. Shaver points, clock points and bell transformers may be neglected for the purpose of load calculation. As a general rule, ten outlets per circuit is about right. Also remember that fluorescent fittings and discharge lamps are rated by their output, and the output must be multiplied by a factor of 1.8 if exact information is not available (*Table 1A, note 2 of the On-Site Guide*).
- Are the switch returns colour identified at both ends?
- Have the switch drops got CPCs? If they have, are they sleeved with green and yellow?
- Are the CPCs correctly terminated?
- Are the switch boxes made of box wood or metal?
- Are ceiling roses suitable for the mass hanging from them?
- Only one flexible cord should come out of each ceiling rose unless they are designed for multiple cords.
- Light fittings in bathrooms must be suitable for the zones in which they are fitted.
- Circuits supplying luminaries fitted outside must have a 0.4 second disconnection time.
- Is the luminaire suitable distance from a combustible surface. (Regulations 422.3.1 & 422.4.2)
- Are the luminaries suitable for the surface to which they are fixed. (Regulation 422.4.2 note 1 & 2)
- Is the phase conductor to ES lampholders connected to the centre pin? This does not apply to E14 and E27 lampholders (*Regulation 612.6*).

## Three phase circuit/systems

These circuits should be inspected for the same defects that you could find in other circuits. In addition to this:

- Are warning labels fitted where the voltage will be higher than expected? For example, a lighting switch with two phases in it, or perhaps where sockets close to each other are on different phases.
- Are conductors in the correct sequence?
- Remember PFC should be double the phase to neutral fault current.

Occasionally other types of circuit will be found, but the same type of inspection should be carried out using common sense.

> Always remember that the reason for this inspection is to ensure safety

## Periodic testing

The level of testing will usually be far less for periodic testing than it is for initial verification; this is providing that previous test results are available. If they are not, then it will be necessary for the full survey and the complete range of tests to be carried out on the installation, to provide a comprehensive set of results.

The level of testing will depend largely on what the inspector discovers during the visual inspection, and the value of results obtained while carrying out sample testing. If any tests show significantly different results, then further testing may be required.

In some cases, up to 100% of the installation will need to be tested. Periodic testing can be dangerous, and due consideration should be given to safety. Persons carrying out the testing must be competent and experienced in the type of installation being tested and the test instruments being used.

There is no set sequence for the testing which may be required for the completion of the periodic inspection report. The sequence and type of tests which are to be carried out are left to the person carrying out the test to decide upon. Where tests are required, the recommendations for these tests would be:

| | Recommended tests |
|---|---|
| *Continuity of protective conductors* | Between the distribution board earth terminal and exposed conductive parts of current using equipment. Earth terminals of socket outlets (test to the fixing screw of outlet for convenience). |
| *Continuity of bonding conductors* | All main bonding and supplementary bonding conductors. |
| *Ring circuit continuity* | Only required where alterations or additions have been made to the ring circuit. |
| *Insulation resistance* | Only between live conductors joined and earthed. Or between live conductors with the functional switch open if testing lighting circuit. |
| *Polarity* | Live polarity tested at the origin of the installation. Socket outlets. At the end of radial circuits. Distribution boards. |

| | |
|---|---|
| *Earth electrode resistance* | Isolate installation and remove earthing conductor to avoid parallel paths. |
| *Earth fault loop impedance* | At the origin of the installation for $Z_e$. Distribution boards for the $Z_e$ of that board. Socket outlets and at the end of radial circuits for $Z_S$. |
| *Functional tests* | RCD tests and manual operation of isolators, protective devices and switches. |

## Voltage drop in conductors

It is part of the inspection process to ensure that installed conductors have been correctly selected for current carrying capacity and voltage drop. To check the suitability of the current carrying capacity it is simply a matter of looking at the installation method, and then checking on the current carrying capacity tables for the cable in Appendix 4 of BS 7671.

To ensure that the cable meets the voltage drop requirements is slightly more complex. A simple method is to measure the voltage at the origin of the circuit, and then measure the voltage at the end of the circuit with the load connected and switched on. The difference between the two measurements will be the volt drop.

If the first method is impractical, then a resistance test should be carried out between the phase and neutral of the circuit. This test is carried out using the same method as the $R_1 + R_2$ test although, instead of the test being between phase and CPC, it is between the phase and neutral for the circuit. Once the resistance $R_1 + R_n$ of the circuit has been measured it should be multiplied by the current that will flow in the circuit. This will give you the volt drop for the circuit.

## Example

A circuit is wired in $2.5\,\text{mm}^2$ and is 25 metres in length. The current in the circuit is 18 amps.

*The measured value of resistance is* $0.37\,\Omega$

$$voltage\ drop = I \times R = V$$
$$18 \times 0.37 = 6.66\ Volts.$$

*This is the voltage drop for the circuit.*

# 3

# Testing of electrical installations

## SAFE ISOLATION

It cannot be over-emphasized how important it is that isolation of electrical circuits is carried out in a set sequence, and that this sequence is repeated each time a circuit or complete installation is to be isolated.

If the same procedure is followed each time isolation is carried out, it will soon become a habit, which can only be a good thing as it may save your life.

It is vital that the correct test equipment is used for isolation and that it complies with the Health and Safety Executive document GS 38. This document gives guidance on the use of test equipment, particularly leads and probes.

The GS 38 document is not a statutory document but if the guidance given in the document is followed, it will normally be enough to comply with the Health and Safety at Work Act 1974, the Electricity at Work Regulations 1989 and any other statutory requirements that may apply. The items of equipment that should be available to persons carrying out the safe isolation procedure are:

• A proving unit

- An approval voltage
  indicator (left) and
  test Lamp (right)

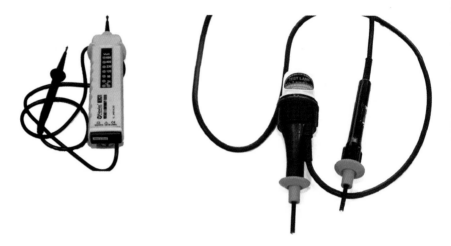

- Warning notices

- Locking devices

Another useful piece of equipment is:

- An $R_1$ and $R_2$ box. This will not only be useful for the safe isolation of socket outlets, it can also be used for ring circuit testing and the $R_1 + R_2$ testing of radial circuits incorporating a socket or socket outlets without having to remove them from the wall.

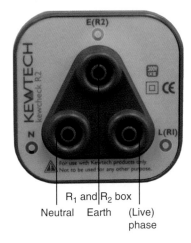

The leads should be:

- Flexible and long enough, but not too long.
- Insulated to suit the voltage at which they are to be used.
- Coloured where it is necessary to identify one lead from the other.
- Undamaged and sheathed to protect them against mechanical damage.

The probes should:

- Have a maximum of 4 mm exposed tip (preferably 2 mm).
- Be fused at 500 mA or have current limiting resistors.
- Have finger guards (*to stop fingers slipping on to live terminals*).
- Be colour identified.

## Isolation procedure

It is very important to ensure that the circuit that you want to isolate is live before you start. To check this, a voltage indicator/test lamp or a piece of equipment that is already connected to the circuit should be used. If it appears that the circuit is already dead, you need to know why.

- Is somebody else working on it?
- Is the circuit faulty?
- Is it connected?
- Has there been a power cut?

You must make absolutely certain that you and you alone are in control of the circuit to be worked on. Providing the circuit is live you can proceed as follows:

### STEP 1

Ensure voltage indicator/test lamp is working correctly.

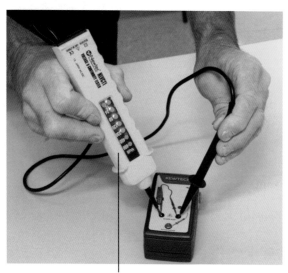

Voltage lights lit

### STEP 2

Test between all live conductors and live conductors and earth.

Voltage lights lit

### STEP 3

Locate the point of isolation. Isolate and lock off.

Place warning notice (**DANGER ELECTRICIAN AT WORK**) at the point of isolation.

Isolate and lock off          Place warning notice

### STEP 4

Test circuit to prove that it is the correct circuit that you have isolated.

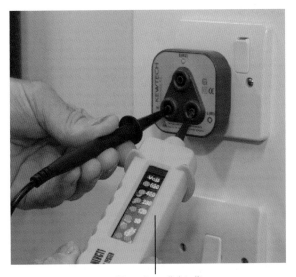

No voltage lights lit

Be careful! Most test lamps will trip an RCD when testing between live and earth, it is better to use an approved voltage indicator to GS 38 as most of these do not trip RCDs

## STEP 5

Check that the voltage indicator is working by testing it on a proving unit or a known live supply.

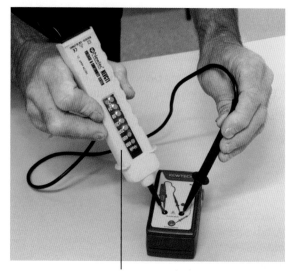

Voltage lights lit

When carrying out the safe isolation procedure never assume anything, always follow the same procedure

It is now safe to begin work.

If the circuit which has been isolated is going to be disconnected at the consumer's unit or distribution board, **REMEMBER** the distribution board should also be isolated. The Electricity at Work Regulations 1989 do not permit live working.

## TESTING OF PROTECTIVE BONDING CONDUCTORS

### Protective equipotential bonding

This test is carried out to ensure that the equipotential bonding conductors are unbroken, and have a resistance low enough to ensure that, under fault conditions, a dangerous potential will not occur between earthed metalwork (*exposed conductive parts*) and other metalwork (*extraneous conductive parts*) in a building.

It is not the purpose of this test to ensure a good earth fault path but to ensure that, in the event of a fault, all exposed and extraneous conductive parts will be live at the same potential, hence EQUIPOTENTIAL bonding. In order to achieve this, it is recommended that the resistance of the bonding conductors does not exceed $0.05\Omega$.

Table 54.8 and Regulation 544.1.2 in BS 7671 cover the requirements of equipotential bonding. Chapter 4 of the *On-Site Guide* is also useful. Maximum lengths of copper bonding conductors before $0.05\Omega$ is exceeded.

| Size mm² | Length in metres |
|----------|------------------|
| 10 | 27 |
| 16 | 43 |
| 25 | 68 |
| 35 | 95 |

The test is carried out with a **Low Resistance Ohm meter** and often can only be carried out on the initial verification; this is because one end of the bonding conductor must be disconnected to avoid parallel paths. When disconnecting a bonding conductor, it is important that the installation is isolated from the supply. On larger installations it is often impossible to isolate the installation and, therefore, the conductor must remain in place. The instrument should be set on the lowest value of $\Omega$ possible.

## STEP 1

Isolate supply (as safe isolation procedure)

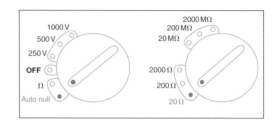

## STEP 2

Disconnect one end of the conductor (*if possible, disconnect the conductor at the consumers unit, and test from the disconnected end and the metalwork close to the bonding conductor. This will test the integrity of the bonding clamp*).

Test leads

## STEP 3

Measure the resistance of test leads or null leads (*these may be long as the only way that we can measure a bonding conductor is from end to end*).

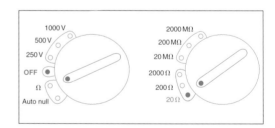

Nulled leads

## STEP 4

Connect one test lead to the disconnected conductor at the consumer's unit.

Note: Safety notice removed for clarity.

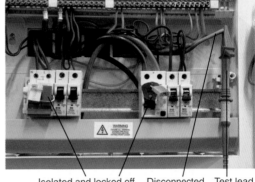

Isolated and locked off    Disconnected    Test lead
conductor

## STEP 5

Connect the other end of the test lead to the metalwork that has been bonded (connecting the lead to the metalwork and not the bonding clamp will prove the integrity of the clamp).

Bended metalwork      Other end of test lead

## STEP 6

If the instrument is not nulled remember to subtract the resistance of the test leads from the total resistance. This will give you the resistance of the bonding conductor. If the meter you are using has been nulled, the reading shown will be the resistance of the conductor.

Very low value—less than 0.05A

## STEP 7

Ensure that the bonding conductor is reconnected on completion of the test.

Whilst carrying out this test a visual inspection can be made to ensure that the correct type of BS 951 earth clamp, complete with label is present, and that the bonding conductor has not been cut if it is bonding more than one service.

If the installation cannot be isolated on a periodic inspection and test, it is still a good idea to carry out the test; the resistance should be a maximum of $0.05\Omega$ as any parallel paths will make the resistance lower. If the resistance is greater than $0.05\Omega$ the bonding should be reported as unsatisfactory and requires improvement.

In some instances the equipotential bonding conductor will be visible for its entire length; if this is the case, a visual inspection would be acceptable, although consideration must be given to its length.

For recording purposes on inspection and test certificates no value is required but verification of its size and suitability is.

Items to be bonded would include any incoming services, such as: water main, gas main, oil supply pipe, LPG supply pipe. Also included would be structural steel work, central heating system, air conditioning, and lightning conductors within an installation (*Bonding to lightning conductors must comply with BS EN 62305. It is advisable to seek advice from a specialist before connecting any bonding to them*).

This is not a concise list and consideration should be given to bonding any metalwork that could introduce a potential within a building.

Correct

Incorrect

## Continuity of supplementary equipotential bonding conductors

There are two general reasons for carrying out supplementary equipotential bonding.

### *Supplementary bonding*

Supplementary bonding is required where circuit disconnection times cannot be met (Regulation 411.3.2.6) or where there is an increased risk of electric shock. Generally, these areas would be in bathrooms, swimming pools and other special locations.

Where disconnection times cannot be met and the effectiveness of supplementary bonding is required to be checked (Regulation 415.2.2) a simple test is required.

Where automatic disconnection is by a protective device, the regulations give a formulae $R \leq 50V/I_a$. If the circuit is protected by an RCD, the formulae becomes $R \leq 50V/I_{\Delta n}$ (in some special locations 25V must be used).

When this formula is used, it will ensure that any touch voltage in the bonded area will not rise above 50 volts a.c. before the protective device operates.

To use the formula where a protective device is in place for automatic disconnection of supply (ADS), the first step is to find $I_a$ for a 5 second disconnection time. Let's say that the device is a 32a BS 88 fuse.

Look in Appendix 3 of BS7671, Table 3.3 A. There is a grid in the top right-hand corner of the page, and it can be seen that for a 32A device a current of 125A is required to operate the fuse. This value can also be found by using the maximum $Z_s$ value for a 5 second disconnection time. The $Z_s$ value can be found in Table 41.4 in part 4 of BS7671. It is $1.84\,\Omega$.

A calculation can now be carried out:

$$I_a = \frac{U_O}{Z_S}$$

$$I_a = \frac{230}{1.84}$$

$$I_a = 125\ A$$

(*This method can be used for all protective devices*)

Now the values can be used to verify that the area does or does not require supplementary bonding to be installed. The calculation is:

$$R = \frac{50v}{I_a}$$

$$R = \frac{50}{125} = 0.4 \, \Omega$$

$0.4 \, \Omega$ is the maximum value permitted between exposed or extraneous conductive parts in the area. If when measured it is found that the resistance is higher than $0.4 \, \Omega$ then supplementary bonding will be required. Clearly, the value of resistance will not be the same for different ratings or types of protective devices.

Where an RCD is installed to protect the circuit the calculation to find out if supplementary bonding is required is as follows:

$$R = \frac{50v}{I_{\Delta N}}$$

$I_{\Delta N}$ is the operating current of the RCD. If it is a 30 mA device the calculation is as follows:

$$R = \frac{50}{0.03}$$

$$R = 1666 \, \Omega$$

Apart from where automatic disconnection times cannot be met and special locations, due consideration must be given to other areas where there is an increased risk of electric shock. There is no specific prerequisite to carry out supplementary bonding in Kitchens, particularly since the introduction of the 17th edition of BS7671 which requires the use of RCD protection for most circuits. However, if it is thought by the installer that there is an increased risk of electric shock, there is no reason why bonding could not be carried out – it will do no harm providing it is carried out correctly.

On occasions it is often useful to carry out supplementary bonding, particularly under Kitchen sinks. This may not be for electrical reasons, but more for visual purposes. Bonding is not well-understood by many people. Let's say that you have travelled 20/30 miles to fit a Kitchen and completed everything to comply with the required regulations. A few days later, however, and before you have been paid for the work, you receive a phone call from your customer, informing you that his next door neighbour has spotted that you have not bonded the sink. Of course your customer will believe that his neighbour is right and that you have forgotten something, or tried to save a bit of money. The choice is now yours – do you try and convince your customer that his neighbour is wrong, or do you travel back to the job to carry out the bonding to ensure payment?

Perhaps for the sake of a couple of earth clamps and a short length of $4\,mm^2$, it would have been cheaper just to use the bonding in the first place.

Where supplementary bonding is used, a test must be carried out to ensure that the resistance between exposed and extraneous conductive parts is in place and has a resistance of less than the value as calculated using the formulae as described earlier. The instrument to be used is known as a low resistance ohm meter.

A visual check must be made to ensure that the correct earth clamps have been used and that they have the correct labels attached.

It is perfectly acceptable to utilize the pipe work and structural steelwork within the area to be used as a bonding conductor, and bonding can be carried out adjacent to the area providing that the integrity of the pipe work/ steelwork can be assured. An airing cupboard would be a good example of a suitable place to bond.

If it is necessary for the lighting point or electric shower in a bathroom, there is no reason why the bonding conductor could not be simply attached to a pipe within the roof space, which is bonded elsewhere and passes near the item which requires bonding. The correct bonding clamps and labels should always be used.

If using the pipe work of plumbing and heating systems as bonding conductors, the continuity of the pipe work must be verified. Tests must be made between exposed and extraneous conductive parts to ensure that the resistance does not exceed the values as calculated using the formulae described earlier. This is a simple test carried out using a low resistance ohm meter. A probe of one lead should be placed on one metal part and the probe of the other lead placed on an adjacent metal part.

Probe on solid metal part of tap

Probe on unpainted metal work

Problems can arise if the pipe work is altered and plastic push fittings are used. Clearly these will not conduct and the bonding continuity could be compromised. If ever a plastic plumbing fitting is used on copper pipe work, consideration should be given to the installation of a bonding conductor installed across the fitting.

Bonding conductor

It is a common belief that water in pipe work will conduct; in fact, the current that would flow through water in a 15 mm diameter pipe which has a plastic joint in it is very small.

To find out just how much, I set up a simple controlled experiment, two short lengths of 15 mm pipe were joined using a 15 mm plastic push fit coupler. The pipe was then filled with water and the two ends of the joined pipe were connected to a 230 volt supply. The current flowing was measured to be 0.003 Amperes (3 mA). The current flow would increase if the water had central heating additives in it, but not considerably.

Exposed conductive parts

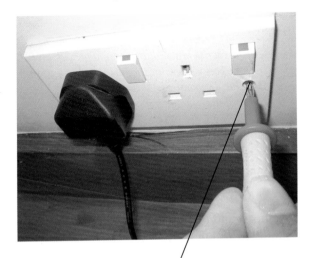

Screw fixings are earthed

Extraneous conductive parts

Unpainted metal work

---

## Example 1

Determining if supplementary bonding is required

A circuit on a TN system is to be altered. The current carrying capacity and the volt drop of the cable is adequate for the load. However, the $Z_s$ value is too high for the 20 amp BS 1361 protective device which is being used to protect the circuit.

The maximum resistance permissible between exposed and extraneous conductive parts must be calculated.

The first step is to find the current that would cause automatic disconnection of the supply.

$$I_a = \frac{U_O}{Z_S} \qquad I_a = \frac{230}{2.80} \qquad = 82.14\,\text{amps}$$

This value can also be found in Figure 3.1 of Appendix 3 of BS 7671 (it is rounded down to 82 amps).

The maximum permissible resistance between conductive parts can now be found by:

$$R = \frac{50}{I_a} \qquad R = \frac{50}{82} \qquad = 0.6\,\Omega$$

*Supplementary bonding is installed where there is a risk of simultaneous contact with any extraneous and exposed conductive parts. Its purpose is to ensure that the potential between any of these parts does not rise above a safe value. In most cases, this value is 50 volts, although some chapters in Part 6 of BS 7671 require a maximum potential of only 25 volts.*

---

*Determining if a part is extraneous, or just a piece of metal*
A test should be made using an insulation resistance tester set on MΩ, supplying 500 volts.

Connect one test lead to the metal part and the other lead to a known earth. If the resistance value is 0.02 MΩ (20,000 Ω) or greater, no supplementary bonding is required. If less than 0.02 MΩ, supplementary bonding should be carried out.

If we use Ohm's law we can see how this works:

$$\frac{V}{R} = I: \qquad \frac{500}{20,000} = 0.025\,A$$

This shows that a current of 25 mA would flow between the conductive parts; this would of course only be 0.012 amp if the fault was on a single phase 230 volt supply. This current is unlikely to give a fatal electrical shock.

The test must not be confused with a continuity test. It is important that an insulation resistance tester is used.

## Continuity of circuit protective conductors

This test is carried out to ensure that the CPC of radial circuits are intact and connected throughout the circuit. The instrument used for this test is a **low resistance ohm meter** set on the lowest value possible.

**This is a dead test and must be carried out on an isolated circuit.**

Testing can be carried out using two methods.

### METHOD ONE

**STEP 1**

Using a short lead with a crocodile clip on each end, bridge phase and CPC together at one end of the circuit (*it does not matter which end, although it is often easier to connect at the distribution board as this will certainly be one end of the circuit*).

The resistance of this lead plus the resistance of the test leads should be subtracted, or the instrument nulled before the $R_1 + R_2$ reading is recorded.

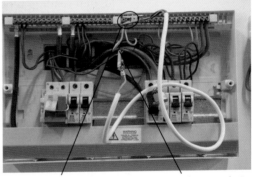

Connect to main earth terminal    Connect to phase conductor

**STEP 2**

At **each** point on the circuit test between phase and CPC.

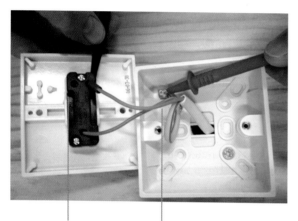

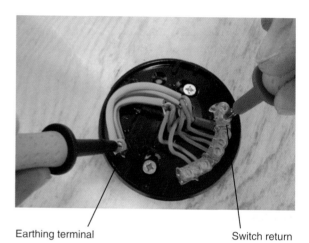

Terminal of switch return    Earthing terminal

Earthing terminal                              Switch return

Keep a note of the readings as you carry out the test, they should increase as you move further from the connected ends. The highest reading obtained should be at the furthest end of the circuit and will be the $R_1 + R_2$ of the circuit. This value should be recorded on the schedule of test results. If the highest reading is obtained at a point which is not the furthest from the circuit, further investigation should be carried out as it may indicate a loose connection (high resistance joint).

In some instances only the value of $R_2$ may be required. Where the phase conductor is the same size as the CPC the total measured resistance can be divided by 2 as the phase and CPC resistance will be the same. If the phase and CPC are of different sizes (this is usual in twin and earth thermoplastic cable) the $R_2$ value can be calculated using the following formula:

$$R_2 = R_1 + R_2 \times \frac{A\ phase}{ACPC + A\ phase}$$

$R_2$ = resistance of CPC in ohms; $R_1 + R_2$ = measured value of resistance in ohms; *A phase* = Area of phase conductor in mm$^2$; *A CPC* = Area of CPC mm$^2$.

## Example 2

A radial circuit is wired in $2.5\,mm^2$ phase and $1.5\,mm^2$ CPC. The test resistance of $R_1 + R_2$ is $0.37\,\Omega$.

*To calculate the resistance of the CPC on its own:*

$$R_2 = 0.37 \times \frac{2.5}{2.5 + 1.5}$$
$$R_2 = 0.37 \times 0.625 = 0.23\,\Omega$$

*If the CPC is smaller than the phase conductor, the resistance of the CPC conductor will always be greater than the phase conductor as it has a smaller cross-sectional area.*

*Another method of determining $R_2$ is described in the ring circuit test.*

### METHOD TWO

This method will prove CPC continuity and is usually only used where the circuit is wired in steel enclosures where parallel paths to the CPC may be present and the $R_1 + R_2$ value would not be a true value. Or where the CPC resistance is required for use with Table 41C of the Wiring Regulations.

This method uses a long lead. One end is connected to the earth terminal of the distribution board, and the other connected to a **low resistance ohm meter**. The short lead of the ohm meter is then touched onto each fitting to ensure that it is connected to the CPC. The highest reading minus the resistance of the leads can be recorded as the $R_2$ reading.

If the furthest point of the circuit is known, and no parallel paths exist, the $R_1 + R_2$ reading can be carried out first using **Method one**, and then a test between earthed metal at each point can be made to ensure that the CPC is connected to each point on the circuit, using **Method two**. This method is particularly useful where there are a lot of enclosed metal fittings and dismantling them would be impractical.

### RING FINAL CIRCUIT TEST

The purpose of this test is to ensure that:

- The cables form a complete ring.
- There are no interconnections.
- The polarity is correct on all socket outlets.

When this test is carried out correctly it also gives you the $R_1$ and $R_2$ value of the ring and identifies spurs.

Appendix 15 in BS7671 provides information regarding final circuits for socket outlets. It states that a ring circuit is to be wired in 2.5 mm$^2$ phase conductor and 1.5 mm$^2$ CPC as a minimum size. This type of circuit should be protected by a 30/32 amp overcurrent protective device.

### Complete ring circuit

A test must be carried out on the conductors to verify that they form a complete loop. If it is found that they do not, overloading of the cables could occur. In installations where more than one ring circuit has been installed, it is possible for the ends of the ring to become muddled, resulting in the circuits being supplied through two protective devices.

The whole point of a ring circuit is that it can be wired in small CSA cables but carry a reasonably high current, this is because we have two 2.5 mm$^2$ cables wired in parallel (*Regulation 433.4*). If we look at Table 4D5 in BS 7671, the value of current that 2.5 mm$^2$ cable can carry is 20 amps in the worst type of conditions.

If we use two of these conductors in parallel, we will have a total current carrying capacity of 40 amps. As one of the jobs of the protective device is to protect the cable, this situation will be fine because the protective device is smaller than the total current carrying capacity of the cables in parallel.

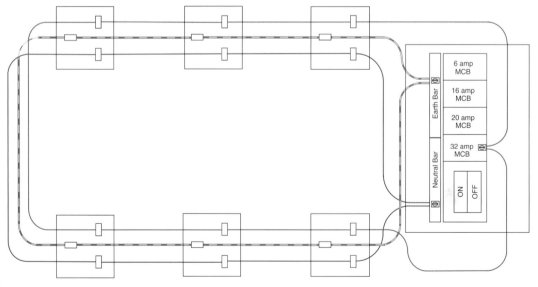

Complete ring circuit

## Broken ring circuit

If, however, we found the ring to be broken, the protective device could not do its job as it is rated at 32 amps and the cable is rated only at 20 amps. Hence overloading!

## Interconnections

Occasionally a situation will be found where there is a ring within a ring, in other words the ring is interconnected.

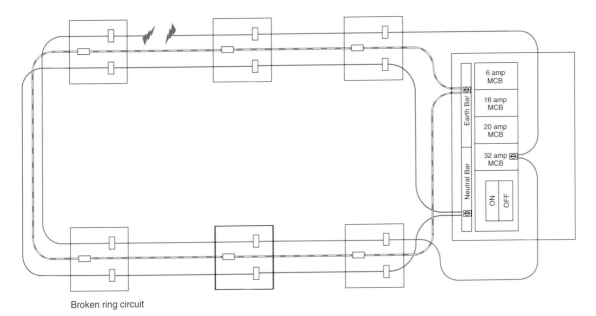

Broken ring circuit

This situation, as it is, will not present a danger. However, it will make it very difficult for a ring final circuit test to be carried out as, even if the correct ends of the ring are connected together, different values will be found at various points of the ring. If one loop is broken, a test at the consumer's unit will still show a complete ring. It will not be until further tests are performed that the interconnection/broken loop will be found.

## Polarity

Each socket outlet must be checked to ensure that the conductors are connected into the correct terminals. Clearly if they are not, serious danger could occur when appliances are plugged in.

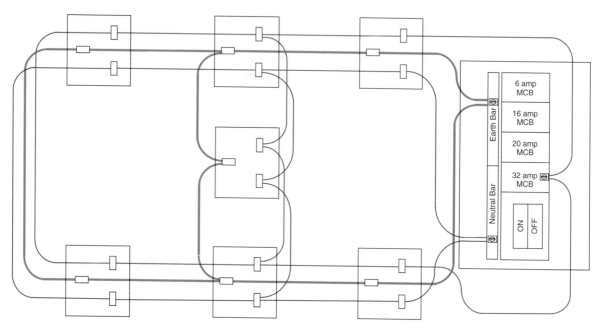

Interconnected ring circuit

It could be that phase and neutral are the wrong polarity; the result of this is that the neutral would be switched in any piece of equipment with a single pole operating switch.

If the live conductors and CPC are connected with reverse polarity, then the case of any Class 1 equipment could become live and result in a fatal electric shock.

### Performing the test

The instrument required is a low resistance ohm meter set on the lowest scale, typically $20\,\Omega$. Be sure to zero the instrument or subtract the resistance of the leads each time you take a reading.

This is a dead test! Safe isolation must be carried out before working on this circuit

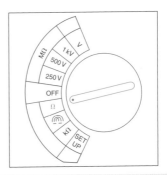

## STEP 1

Isolate circuit to be tested.

## STEP 2

Identify legs of ring.

## STEP 3

Test between ends of phase conductor and note the resistance value.

Instrument set to Ω for whole test.

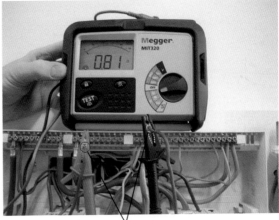

Ends of phase conductor

## STEP 4

Test between ends of neutral conductor. This value should be the same as the phase conductor resistance as the conductor must be the same size (*see Note 1*).

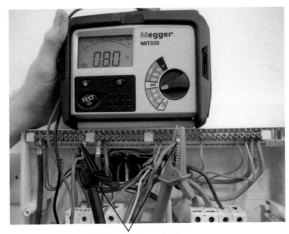

Ends of neutral conductor

### STEP 5

Test between the ends of the CPCs. If the conductor size is smaller than the live conductors (*as is usually the case when using twin and earth cable*), the resistance value will be higher (*see Note 2*); make a note of this reading.

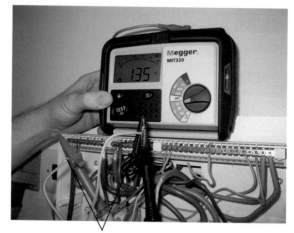

Ends of CPCs

### STEP 6

Join P of leg 1 to N of leg 2.

Test between N of leg 1 and P of leg 2. The measured resistance should be double that of the phase conductor.

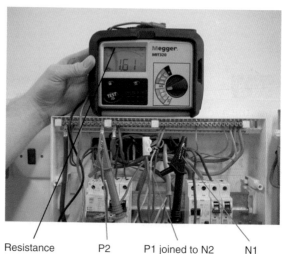

Resistance          P2          P1 joined to N2          N1
double that of
phase conductor

### STEP 7

Join N of leg 1 to P of leg 2 together (*leaving N2 and P1 joined*).

Test between joined ends.

The measured value should be ¼ of test between N of leg 1 and P of leg 2.

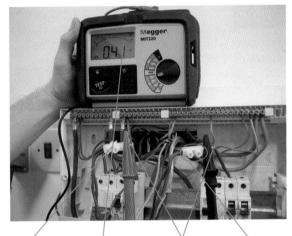

P2 joined
to N1

Resistance 1/4
of that tested
between
N1 and P2

Test between
joined ends

P1 still joined
to N2

### STEP 8

Leave the ends joined.

Test between P and N at each socket outlet, the resistance should be the same at each socket (*see Note 1*).

A higher reading should be investigated, although it will probably be a spur it should be checked as it may be a loose connection (*high resistance joint*).

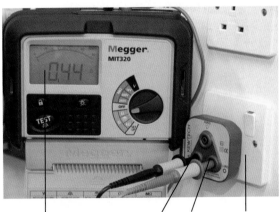

Resistance value
the same at each
socket

N      P

Test at
each socket
outlet

**STEP 9**

Disconnect the ends and repeat the test using phase and CPC conductors (*see Note 3*).

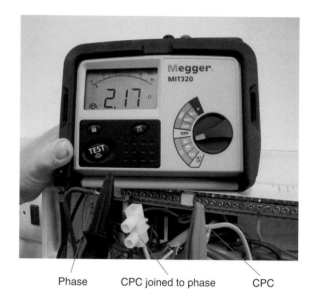

Phase    CPC joined to phase    CPC

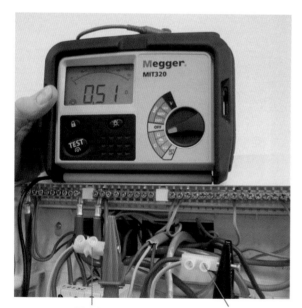

P2 joined to CPC 1    P1 joined to CPC 2

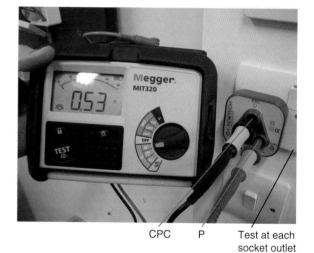

CPC    P    Test at each socket outlet

The highest value (*which will be the spur*) will be the $R_1$ and $R_2$ value for this circuit.

Notes

1. If with ends connected (P1/N2 and P2/N1) a substantially different resistance value is measured at each socket outlet, check that the correct ends of ring are connected. A difference of 0.05Ω higher or lower would be acceptable.
2. In a twin and earth cable the CPC will usually have a resistance of 1.67 times that of the phase conductor as it has a smaller cross-sectional area.
3. When phase and CPC conductors are not the same size a higher resistance value will be measured between Phase and CPC than Phase and neutral. It will also alter **slightly** as the measurement is taken around the ring, the resistance will be lower nearer the joined ends and will increase towards the centre of the ring. The centre socket of the ring will have the same resistance value as the test between the joined ends.
4. If the circuit is contained in steel conduit or trunking parallel paths may be present, this would result in much lower $R_1 + R_2$ resistance values.
5. Some certificates may require $r_n$ to be documented. This is the resistance of the neutral loop measured from end to end.

## Example 3

Let's use a 2.5/1.5 mm$^2$ twin and earth cable 22 metres long. If we look in Table 9A in the *On-Site Guide* we will see that the resistance of a copper 2.5 mm$^2$ conductor has a resistance of 7.41 mΩ per metre.

*The resistance of the phase conductor will be:*

$$\frac{7.41 \times 22}{1000} = 0.163\,\Omega$$

*Divide the largest conductor by the smallest to find the ratio of the conductors (how much bigger is the larger conductor?).*

$$\frac{2.5}{1.5} = 1.67$$

*The 2.5 mm$^2$ conductor is 1.67 × larger than the 1.5 mm$^2$ conductor; therefore, it must have 1.67 × less resistance than the 1.5 mm$^2$ conductor.*

*If we now multiply the resistance of the phase conductor by 1.67:*
*0.163 × 1.67 = 0.27 Ω this is the resistance of the 1.5 mm² conductor.*

*We can check this by looking at Table 9A of the On-Site Guide once again, and we can see that the resistance of 1.5 mm² copper is 12.10 mΩ per metre. Therefore, 22 metres of 1.5 mm² copper will be:*

$$\frac{22 \times 12.10}{1000} = 0.266 \; \Omega$$

*As a final check, if we look at Table 9A of the On-Site Guide for the resistance of a 2.5 mm²/1.5 mm² cable, we will see that it has a resistance of 19.51 mΩ per metre, and that 22 metres of it will have a resistance of:*

$$\frac{22 \times 19.51}{1000} = 0.429 \; \Omega$$

*The resistance value of the 2.5 mm² is 0.163 Ω; and the resistance value of the 1.5 mm² is 0.266 Ω.*

*If we add them together: 0.163 Ω + 0.266 Ω = 0.429 Ω. Finally, 0.429 Ω is the resistance of our 2.5 mm²/1.5 mm² measured as one cable.*

---

### INSULATION RESISTANCE TEST

This is a test that can be carried out on a complete installation or a single circuit, whichever is suitable or required. The test is necessary to find out if there is likely to be any leakage of current through the insulated parts of the installation. A leakage could occur for various reasons.

A good way to think of this test is to relate it to a pressure test – we know that voltage is the pressure where the current is located in a cable. On a low voltage circuit, the expected voltage would be around 230V a.c. The voltage used in an insulation test on a 230V circuit is 500V, which is more than double the normal circuit voltage. Therefore, it can be seen as a pressure test similar to a plumber pressure testing the central heating pipes.

### *Low insulation resistance*

Cable insulation could deteriorate through age. A low insulation resistance caused through age will often be found in installations where rubber-insulated cables have been used. Cables which are crushed under floor boards, clipped on edge, or worn thin where pulled through holes in joists next to other cables, can give a very low reading.

Low insulation resistance could be found if a building has been unused for a period of time, due to the installation being affected by dampness in the accessories. Low insulation resistance readings will also often be found where a building has been recently plastered. In theory, long lengths of cables or circuits in parallel could give low readings due to the amount of insulation (*the longer the circuit or the more circuits, the more insulation there will be for leakage to occur*).

The instrument used to carry out this test is an **insulation resistance tester**. To comply with the requirements of the Health and Safety Executive the instrument must be capable of delivering a current of 1 mA applied to a resistance of 1 MΩ. Table 61 in BS7671 gives the test voltages and minimum acceptable resistance values.

The values are:

For circuits between 0 volts and 50 volts a.c., a test voltage of 250 volts d.c. is required and the minimum acceptable value is given as 0.5 MΩ.

The values are shown here:

| | Circuits between 0 V and 50 V a.c. | Circuits between 50 V a.c. and 500 V a.c. | Circuits between 500 V and 1000 V a.c. |
|---|---|---|---|
| Required test voltage | 250 V d.c. | 500 V d.c. | 1000 V d.c. |
| In 17th Edition of the Wiring Regulations | (0.5 MΩ) | (1 MΩ) | (1 MΩ) |

### Domestic installations

Remember that testing should be carried out from the day the installation commences (Regulation 610.1).

### Testing a whole installation

In <u>new</u> domestic installations it is often easier to carry out insulation resistance testing on the whole of the installation from the meter tails before they are connected to the supply. If this is the preferred choice the test should be carried out as follows:

- Safe isolation must be carried out before commencing this test.
- Inform any occupants of the building that testing is to be carried out.
- Ensure that all protective devices are in place and switched on.
- Remove all lamps from fittings where accessible.
- If the lamps are not accessible or if a luminaire with control gear (fluorescent) is connected, open the switch controlling the luminaire (*Note 6*).

The same applies to extra low voltage transformers.

- Where dimmer switches are fitted it is important that they are either removed and the switch wires joined, or that the switch is bypassed (*Note 7*).
- Any accessories with indicator lamps are switched off (*Note 8*).
- Passive infra red detectors (PIRs) are removed or bypassed (*Note 9*).
- All fixed equipment such as cookers, immersion heaters, boilers and television amplifiers are isolated.
- Shaver sockets are disconnected or isolated (*Note 9*).
- Items of portable equipment are unplugged.

Great care must be taken as, during this test, 500V will be passed through any electrical equipment which is left connected. This could damage the equipment or, at the very least, cause low readings to be obtained during the test. Once all precautions have been taken proceed with the test as follows:

### Notes

1. The control equipment within discharge lamps will cause very low readings. It is quite acceptable to isolate the fitting by turning off the switch. This is more desirable than disconnecting the fitting. After the test between live conductors is completed the control switch for the luminaire should be closed before carrying out the test between live conductors and CPC. This is to ensure that all live conductors are tested for insulation resistance to earth.
2. Most dimmer switches have electronic components in them and these could be damaged if 500V were to be applied to them. It is important that wherever possible the dimmer switches are removed and the phase and switch return are joined together for the test.
3. Neon indicator lamps will be recognized as a load by the test instrument and will give a very low insulation value. All that is required is for the switch on the accessory to be turned off.
4. Passive infra red detectors will give very low readings and may be damaged by the test voltage. Either disconnect it or test between live conductors and earth only on circuits containing PIRs. The same applies to shaver sockets.

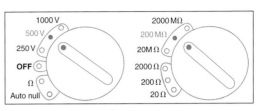

## STEP 1

Set insulation resistance tester to 500V.

Some instruments have settings for Meg ohms and some are self ranging. If yours requires setting then 200 MΩ or higher is the setting to use.

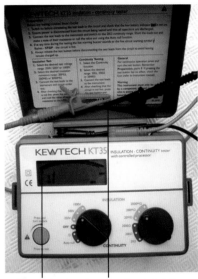

## STEP 2

To ensure that the test results are accurate it is important to ensure correct operation of instrument and the integrity of the leads. Push the test button with the leads disconnected. The resistance shown on screen should be the highest that the instrument can measure.

Over range    Leads not connected

## STEP 3

Join leads and operate instrument again, the resistance shown on screen should be the lowest value possible (0.0 MΩ) in all cases.

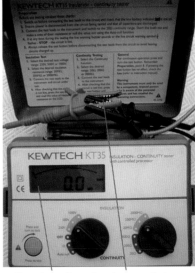

Closed circuit    Leads connected

## STEP 4

When testing the whole installation from the disconnected tails, it is important that the main switch is in the on position and that the protective devices are in place. If they are circuit breakers they must be in the 'on' position.

Test between live conductors (tails) and operate any two-way and intermediate switching. This is to ensure that all switch wires and strappers are tested and that the switch returns have been correctly identified and connected (*no neutrals in the switches*).

Connected to live conductors

Main switch and circuits on

## STEP 5

Join live conductors (tails) together, connect IR tester leads, one on live conductors and the other on the earthing conductor, carry out the test and again operate all two-way and intermediate switching.

Earthing conductor

Live conductors joined together

Main switch and circuits on

## Polarity test on a lighting circuit

### STEP 1

At the origin of the circuit connect the phase and CPC, this can be done with a short lead with crocodile clips at each end.

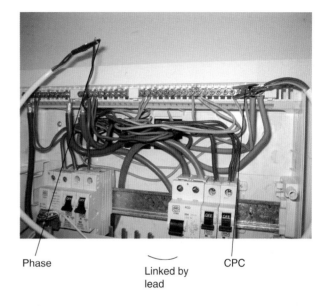

Phase  Linked by lead  CPC

### STEP 2

At the ceiling rose or light fitting place the probes of the instrument on to the earthing terminal and the switched live.

Test instrument probe at earthing terminal  Connection to light fitting place  Test instrument probe at switched live

### STEP 3

Close the switch controlling the light and the instrument should read a very low resistance (this will also be the $R_1$ and $R_2$ reading for the circuit). When the switch is opened the instrument reading should be very high.

This test can also be carried out at the switch if required:

## STEP 1

Place a link between the phase and CPC of the circuit.

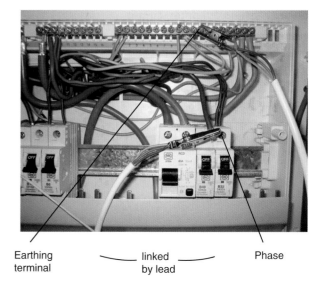

Earthing terminal     linked by lead     Phase

## STEP 2

Place the probes of the test instrument on the earth terminal at the switch and the switch return terminal.

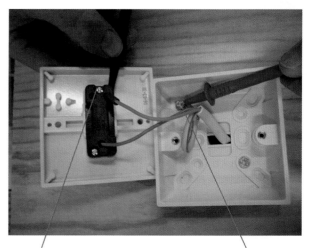

Test instrument probe at switch return terminal     Connection to switch     Test instrument probe at earth terminal

## STEP 3

Close the switch and a low resistance reading
should be shown on the instrument.

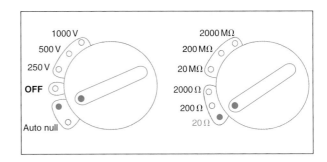

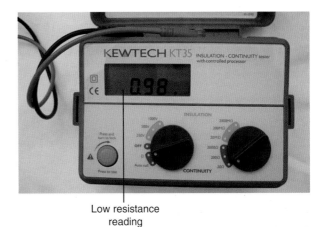

Low resistance
reading

## STEP 4

Open the switch and the instrument reading
will show over range as the circuit should be
open circuit.

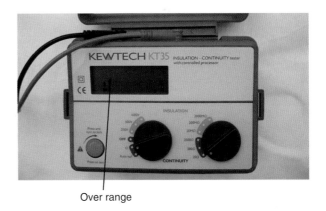

Over range

> Great care must be taken whilst carrying out this test as it is a live test.

### Live polarity test

This test is usually carried out at the origin of the installation before it is energized to ensure that the supply is being delivered to the installation at the correct polarity.

The instrument to be used is an approved voltage indicator or test lamp that complies with HSE document GS 38. It is acceptable for an earth loop impedance meter to be used as these instruments also show polarity.

## STEP 1

Place the probes of the voltage indicator onto the phase and neutral terminal of the incoming supply at the main switch. The device should indicate a live supply.

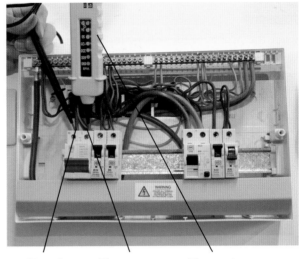

Neutral on main switch

Phase on main switch

Live supply showing

## STEP 2

Place the probes of the voltage indicator onto the phase and earth terminal of incoming supply at the main switch. The device should indicate a live supply.

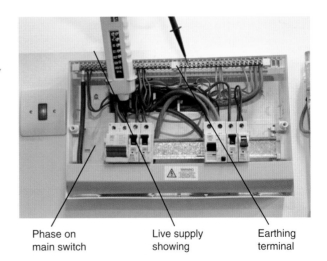

Phase on main switch

Live supply showing

Earthing terminal

### STEP 3

Place the probes of the voltage indicator onto the earthing terminal and the neutral terminal at the main switch. The device should indicate no supply.

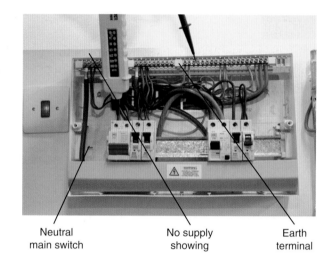

Neutral
main switch

No supply
showing

Earth
terminal

EARTH ELECTRODE TESTING

## Earth fault loop impedance tester

For many installations the resistance of the earth electrode can be measured using an earth fault loop impedance test instrument. It is perfectly acceptable to use this type of instrument on a TT system where reasonably high resistance values could be expected.

The test is performed in exactly the same way as the external earth fault loop $Z_e$ test.

**STEP 1**

Isolate the installation.

**STEP 2**

Ensure that the earthing conductor is correctly terminated at the earth electrode.

**STEP 3**

Disconnect the earthing conductor from the main earthing terminal.

**STEP 4**

Connect a lead of the earth fault loop meter to the disconnected earthing terminal.

**STEP 5**

Place the probe of the other lead on to the incoming phase conductor at the supply side of the main switch and carry out the test.

**STEP 6**

Record the result.

**STEP 7**

Reconnect the earthing conductor and leave the installation in safe working condition.

If a three lead test instrument is used, read the instrument instructions before carrying out the test. It may be that the leads must be connected on to the phase, neutral and earthing terminals, or possibly the neutral and earth lead of the test instrument, should be joined together.

Table 41.5 from BS7671 gives maximum earth fault loop impedances permissible for the correct operation of RCDs.

The values can also be found by the use of the following calculation:

50 is the maximum voltage
$I\Delta n$ is the trip rating of the residual current device
$Z_S$ is the earth fault loop impedance

If the rating of the device is 100 mA the calculation is:

$$\frac{50}{0.1} = 500 \ \Omega$$

The maximum permissible value to comply with the regulations using this calculation is 500 Ω. Although this value would be deemed acceptable, it may not be reliable as it could rise to an unacceptable value if the soil dries out.

An acceptable value for RCDs up to 100 mA is stated as a maximum of 200 Ω. If the resistance is above this it should not be accepted under most circumstances. The installation of an additional or larger electrode may bring the resistance value down to an acceptable value.

The maximum calculated values for earth electrodes are:

| Operating current of the RCD | Electrode resistance in ohms |
|---|---|
| 100 mA | 500 |
| 300 mA | 160 |
| 500 mA | 100 |

For special locations where the maximum touch voltage is 25V the electrode resistance should be halved. Electrode tests should be carried out in the worst possible conditions. The worst condition for an earth electrode is when the soil is **dry**. Where lower values of earth electrode resistances are required, an earth electrode tester should be used.

## Measurement using an earth electrode test instrument

This test requires the use of three electrodes: the earthing electrode under test, a current electrode and a potential electrode.

**STEP 1** The earthing electrode (E) should be driven into the ground in the position that it is to be used. Attention should be paid to the length of the electrode which is in the ground.

**STEP 2** The current electrode ($C^2$ spike) should be pushed into the ground at a distance of ten times the depth of the electrode under test away from it.

**STEP 3** The potential electrode ($P^2$) should be pushed into the ground midway between E and $C^2$.

**STEP 4** The leads of the test instrument should be connected to the appropriate electrodes.

**STEP 5** Measure the value of resistance.

**STEP 6** Move $P^2$ 10% closer to $C^2$.

**STEP 7** Measure the value of resistance.

**STEP 8** Move $P^2$ back to 10% closer to E than the mid-point.

**STEP 9** Measure the value of the resistance.

A calculation must now be carried out to find the percentage deviation of the resistance values.

## Example 6

Three measurements are taken: 79 ohms 85 ohms and 80 ohms.

These must now be added together and an average value calculated.

Total value of three readings is 245 ohms

Find average $\dfrac{244}{3} = 81.33$

The average value is 81.33 ohms

Now find the difference between the average value and the highest measured value. In this case it will be $85 - 81.33 = 3.67$

The percentage of this value to the average value must now be found.

$$\frac{3.67 \times 100}{81.33} = 4.51$$

This value in known as the percentage deviation.

Guidance note 3 tells us that the accuracy of this measurement is typically 1.2 times the percentage deviation. Therefore, to ensure that we use the correct value we must now multiply the percentage deviation by 1.2.

$$4.51 \times 1.2 = 5.41$$

This value is higher than 5% of the average value and it is not advisable to accept it as a percentage deviation if greater than 5% is deemed to be inaccurate. To overcome this, the distance between the electrode under test (E) and the current spike ($C^2$) should now be increased and the first three tests repeated to obtain a more accurate reading.

---

If the required resistance value cannot be obtained by the use of a single electrode, additional electrodes may be added at a distance from the first electrode equal to its depth.

### EARTH FAULT LOOP IMPEDANCE $Z_E$

**This is a live test and great care must be taken**

$Z_e$ is a measurement of the external earth fault impedance (resistance) of the installation. In other words, it is the measured resistance of the supply transformer winding, the supply phase conductor, and the earth return path of the supply. It is measured in ohms.

## Earth fault path for a TT system

This system uses the mass of earth for the fault return path.

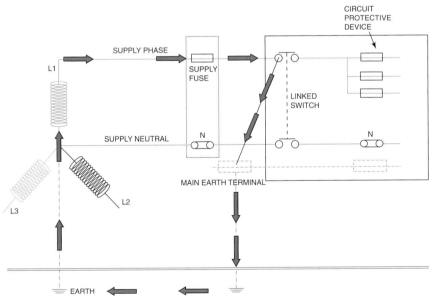

### Earth fault path for a TNS system

This system uses the sheath of the supply cable for the earth fault return path.

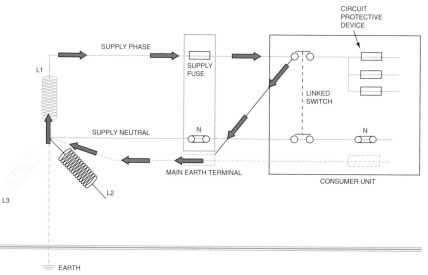

## Earth fault path for a TNCS system

This system uses the neutral (PEN) conductor of the supply for its earth fault
return path.

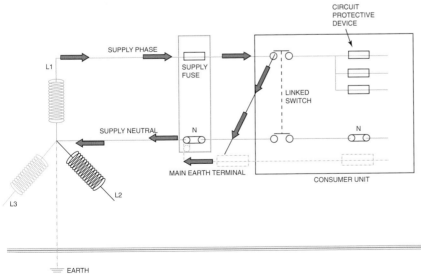

To carry out this test correctly the installation should be isolated from the supply and the main earth disconnected from the main earth terminal (MET). This is to avoid the possibility of parallel paths through any earthed metalwork within the installation.

Often in industrial and commercial installations – where isolation may be impossible due to the building being in use – the only time that this test can be carried out with the main earthing conductor disconnected, is during the initial verification. It is important that the $Z_e$ with the earthing conductor is disconnected during the initial verification as this will give a reference value for the life of the installation. If, during subsequent tests, the earthing conductor cannot be disconnected, a test can still be carried out but the parallel paths should give a lower impedance value. If a higher value is recorded it will indicate a deterioration of the supply earth.

The instrument used for this test is an earth fault loop impedance meter, and it is important that the person using the instrument has read and understood the operating instructions. There are many types of test instruments on the market and they all have their own characteristics.

Some instruments have three leads which must be connected to enable this test to be carried out correctly. Some instruments require that the leads are connected to the phase, neutral and earth of the circuit to be tested. Other instruments require the phase lead to be connected to the phase conductor and the earth and neutral leads to be connected to the earthing conductor of the circuit to be tested (*it is important to read the instructions of the instrument being used*).

If using a two-lead instrument it should be set on $Z_e$. One lead should be connected to the main earthing conductor and the other to the incoming phase on the supply side of the main switch.

Probes can be used for this providing they meet the requirements of GS 38, i.e.:

- Insulated
- Fused
- Finger guards
- Maximum of 4 mm exposed tips or retractable shrouds
- Long enough to carry out test safely
- Undamaged

*Performing the test*

| STEP 1 | Isolate the supply. |
| STEP 2 | Disconnect the earthing conductor. |
| STEP 3 | Set the instrument to loop test. |

> This is a live test and care should be taken when carrying it out

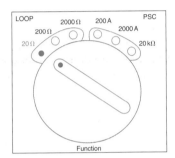

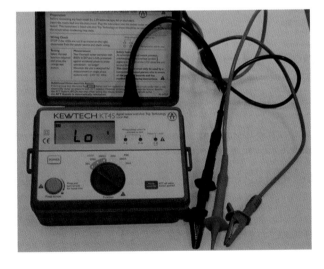

**STEP 4** If you are using a two-lead instrument the leads should be connected as in the figures below. If you are using a three lead instrument, then the leads should be connected as shown here.

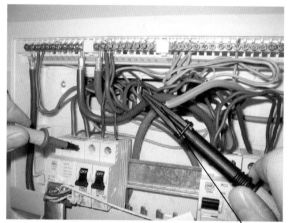

Connections for 2-lead instrument    Disconnected earthing conductor

Connections for 3-lead instrument    Leads joined

**STEP 5** The measurement obtained is $Z_e$ and can be entered on the test certificates in the appropriate place.

**STEP 6** Reconnect earthing conductor.

It is important that the instructions for the test instruments are read and fully understood before carrying out this test.

Care should be taken to reset time clocks, programmers, etc. when the supply is reinstated.

CIRCUIT EARTH FAULT LOOP IMPEDANCE

**This is a live test and great care must be taken**

$Z_S$ is the value of the earth loop impedance (resistance) of a final circuit including the supply cable.

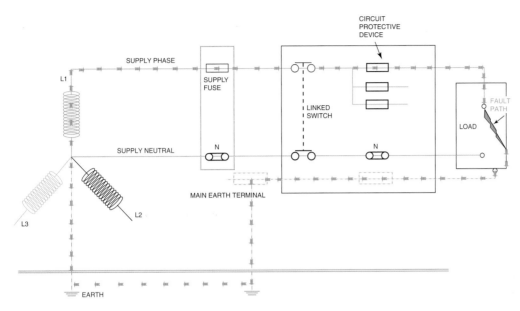

The earth fault loop ($Z_S$) path for a TT system.

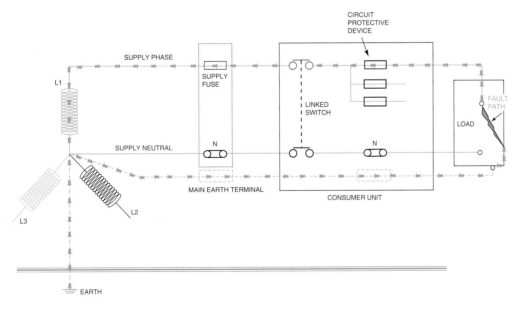

The earth fault loop ($Z_S$) path for a TNS system.

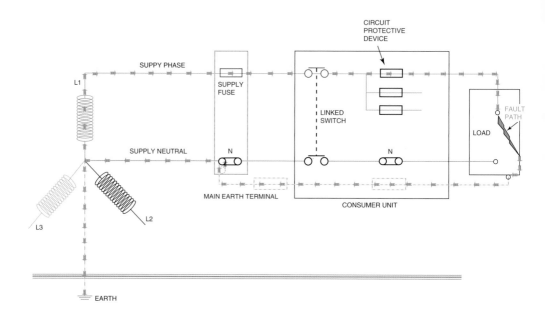

The earth fault loop ($Z_S$) path for a TNCS system.

To obtain $Z_S$, the $Z_e$ value should now be added to the $R_1 + R_2$ values that were obtained when carrying out the CPC continuity tests for each individual circuit.

The total value $Z_e + R_1 + R_2$ will be $Z_S$ (earth loop impedance for the circuit).

This value ($Z_S$) should now be compared with the maximum values of $Z_S$ given in BS 7671, Chapter 41, to verify that the protective device will operate in the correct time.

Unfortunately, it is not quite as simple as it seems. This is because the values $Z_S$ have been measured when the conductors were at room temperature and the maximum $Z_S$ values given in BS 7671 are at the conductor operating temperature of 70°C. This is the maximum temperature that the conductor could be operating at in a sound circuit. There are two methods.

Accurate test instruments must be used for these tests.

METHOD ONE

Measure the ambient temperature of the room and use the values from Table 9B in the *On-Site Guide* as dividers (**DO NOT USE THEM AS MULTIPLIERS**).

This is because Table 9B is to correct the conductor resistances in Table 9A from 20°C to room temperature.

When they are used as dividers they will correct the cable from room temperature resistance, to the resistance that it would be at 20°C.

## Example 7

$$\frac{R_1 + R_2}{\text{Temp factor}}$$

Measured $R_1 + R_2 = 0.84\,\Omega$ @ 25°C

Factor from Table 9b for 25°C = 1.02.

$$\frac{0.84}{1.02} = 0.82$$

Values at 20°C is $0.82\,\Omega$.

*Having corrected the measured values to 20°C the next step is to calculate what the resistance of the cable would be at its operating temperature.*

*As the resistance of copper changes by 2% for each 5°C the conductor resistance will rise by 20% if its temperature rises to 70°C.*

*Multiplying the resistance by 1.2 will increase its value by 20%, this value can now be added to $Z_e$ to give $Z_S$.*

*Resistance value at operating temperature would be:*

$$0.82\,\Omega \times 1.2 = 0.98\,\Omega$$

## Example 8

The $Z_e$ of an installation is $0.6\,\Omega$. A circuit has been installed using twin and CPC 70°C thermoplastic (pvc) cable. The room temperature is 25°C and the measured $R_1 + R_2$ value is $0.48\,\Omega$. The circuit is protected by a BS EN 60898 16A type B device.

Correct the cable resistance to 20°C by using factor from Table 9B *On-Site Guide.*

$$\frac{0.48}{1.02} = 0.47\,\Omega$$

*Adjust this value to conductor operating temperature by increasing it by 20%.*

$$0.47 \times 1.2 = 0.56\,\Omega$$

*Add this value to installation $Z_e$ to find $Z_S$.*

$$0.56 + 0.6 = 1.16\,\Omega$$

*This value can now be compared directly with the maximum value $Z_S$ for a 16A type B protective device. This value is $2.87\,\Omega$ and can be found in Table 41.3 in BS 7671.*

*To comply with the regulations the actual value $1.16\,\Omega$ is acceptable as it is less than $2.87\,\Omega$.*

METHOD TWO

Measure the resistance of earth fault loop impedance at the furthest point of the circuit using the correct instrument (*remember the furthest point is the end of the circuit, not necessarily the furthest distance from the distribution board*). Record the value obtained onto the test result schedule.

This measurement cannot be compared directly with the values from BS 7671 because the operating temperature of the conductors and the ambient temperature of the room are unknown.

This method is useful for a periodic test where existing test results are available. If the measured value is higher than previous results it will indicate that there is a possible deterioration of the earth loop impedance of the circuit.

The usual method to check that the measured $Z_S$ is acceptable is to use the rule of thumb method.

First look in the correct table in Chapter 41 of BS 7671 for the maximum permissible $Z_S$ of the protective device for the circuit being tested.

Use 80% of this value and compare it with the measured value. Providing the measured value is the lowest the circuit will comply.

More information on this can be found in Chapter 5 (Protective devices).

### EARTH LOOP IMPEDANCE USING A HIGH CURRENT LOOP TEST INSTRUMENT WITHOUT TRIPPING AN RCD

It can be very inconvenient if an RCD is tripped by accident. Most electricians will have experience of tripping an RCD, which is being used to protect the whole installation, while using a D lock or low current instrument. This can be very embarrassing and very inconvenient, particularly if it requires the resetting of time clocks and other electronic equipment. Some electricians may not have either of these instruments and will need to rely on the calculation $Z_s = Z_e + R_1 + R_2$. This is fine and perfectly acceptable.

Sometimes it is more satisfying to carry out a live test that will give a direct reading of $Z_s$ to ensure that no loose connections or high resistance joints are affecting the circuit. Some electricians prefer the simplicity of a live test as it leaves very little to chance.

This is a very simple process and it can be carried out as follows:

- Isolate the circuit to be tested.
- Link phase and earth at the furthest point of the circuit using a lead with a crocodile clip on each end, if it is a socket outlet then a plug top with earth and phase linked can be used (*it is advisable to clearly mark the plug top*).
- Use a high current earth fault loop impedance test instrument.
- Place one probe (*black*) on to the isolated terminal of the circuit protective device.
- Place the other probe (*red*) on to the incoming phase of the RCD or main switch.

- Operate instrument and record the result.
- This will be $Z_s$ for the circuit and the RCD will not have tripped.

Performing the tests on non-RCD protected circuits or when using a low current test instrument

If your test instrument is a three lead instrument connect the black and green leads together on to the isolated terminal.

**These are live tests and great care must be taken**

*A circuit incorporating a socket outlet on a ring or a radial*

**STEP 1**  Use an earth fault loop impedance instrument. Set it onto 20 Ω (unless you have a self ranging instrument).

**STEP 2**  Ensure all earthing and bonding is connected.

**STEP 3**  Plug in the instrument and record the reading.

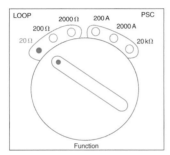

*Performing the test on a radial circuit other than a socket outlet*

### STEP 1

Ensure earthing and bonding is connected.

### STEP 2

Isolate circuit to be tested.

### STEP 3

Remove accessory at the extremity of the circuit
to be tested.

### STEP 4

Use an earth fault loop impedance instrument with
fly leads.

Place the leads on correct terminals. If you are
using a two-lead instrument as connected here;
if you are using a three-lead instrument as
connected here (always read the instrument
manufacturers instructions).

### STEP 5

Energize the circuit.

Probes on phase and earthing
terminals for 2-lead instrument

Two-lead instrument as connected here

Probes on phase and neutral
terminals, clip on earthing terminal
for 3-lead instrument

Three-lead instrumented as connected here

## STEP 2

Place the probes on the phase and neutral terminals at supply side of the main switch.

## STEP 3

Operate the test button and record the reading.

When carrying out the test using a three-lead instrument with leads to GS 38, it is important that the instrument instructions are read and fully understood before carrying out this test.

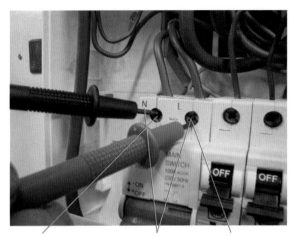

Probe at neutral at supply side of main switch — Insulated tips — Probe at phase at supply side of main switch

Connection for 2-lead instrument

## STEP 1

Place the phase lead on the supply side of the main switch and the neutral and earthing probes/ clips onto the earthing terminal.

Joined leads at neutral at supply side of main switch — Probe at phase at supply side of main switch

Connection for 3-lead instrument

## STEP 2

Operate the test button and record the reading.

If the supply system is a three-phase and neutral system then the highest current that could flow in it will be between phases. Some instruments will not be able to measure the high current that would flow under these circumstances.

Under these circumstances the measurement should be made between any phase and neutral at the main switch and the measured value should be **doubled**.

For your personal safety and the protection of your test equipment it is important to read and fully understand the instructions of your test instrument before commencing this test.

Some PSCC (protective short-circuit current) instruments give the measured value in ohms, not kA. If this is the case, a simple calculation, using ohms law is all that is required.

## Example 14

Measured value is $0.08\,\Omega$.

*Remember to use $U_{oc}$ in this calculation (230 V)*

$$PSCC = \frac{230}{0.08} = 2875\,A$$

It is important that the short circuit capacity of any protective devices fitted exceeds the maximum current that could flow at the point at which they are fitted.

When a measurement of PFC is taken as close to the supply intake as possible, and all protective devices fitted in the installation have a short circuit capacity that is higher than the measured value, then Regulation 432.3 will be satisfied.

In a large installation where sub mains are used to supply distribution boards it can be cost effective to measure the PFC at each board. The PFC will be smaller and could allow the use of a protective device with a lower short circuit rating. These will usually be less expensive.

Table 7.4 in the *On-Site Guide* gives rated short circuit capacities for devices. These values can also be obtained from manufacturer's literature.

| Examples | Rated short circuit capacity |
|---|---|
| Semi-enclosed BS 3036 | 1 kA to 4 kA depending on type |
| BS 1361 Type 1 | 16.5 kA |
| Type 2 | 33 kA |
| BS 88-2.1 | 50 kA at 415 volts |
| BS 88-6 | 16.5 kA at 240 volts |
| | 80 kA at 415 volts |

Circuit breakers to BS 3871 are marked with values M1 to M9 the number indicates the maximum value of kA that they are rated at.

Circuit breakers to BS EN 60898 and RCBOs to BS EN 61009 show two values in boxes, usually on the front of the device.

    Type      In rating     Ics rating     Icn rating

Circuit breaches to BS EN 60898

The square box will indicate the maximum current that the device could interrupt and still be reset $\boxed{3}$. This is the Ics rating. The rectangular box will indicate the maximum current that the device can interrupt safely $\boxed{6000}$. This is the Icn rating.

If a value of fault current above the rated Isc rating of the device were to flow in the circuit, the device will no longer be serviceable and will have to be replaced. A value of fault current above the Icn rating would be very dangerous and possibly result in an explosion causing major damage to the distribution board/consumer's unit.

### FUNCTIONAL TESTING

All equipment must be tested to ensure that it operates correctly. All switches, isolators and circuit breakers must be manually operated to ensure that they function correctly, also that they have been correctly installed and adjusted where adjustment is required.

### Residual current device (RCD)

The instrument used for this test is an RCD tester, and it measures the time it takes for the RCD to interrupt the supply of current flowing through it. The value of measurement is either in seconds or milliseconds.

Before we get on to testing, let's consider what types of RCDs there are, what they are used for, and where they should be used.

### Types of RCD

#### Voltage operated

Voltage operated earth leakage current breakers (ELCBs) are not uncommon in older installations. This type of device became obsolete in the early 1980s and must <u>not</u> be installed in a new installation or alteration as they are no longer recognized by BS 7671.

They are easily recognized as they have two earth connections, one for the earth electrode and the other for the installation earthing conductor. The major problem with voltage operated devices is that a parallel path in the system will probably stop it from operating.

These types of devices would normally have been used as earth fault protection in a TT system.

Although the Electrical Wiring Regulations BS 7671 cannot insist that all of these devices are changed, if you have to carry out work on a system which has one it <u>must</u> be replaced to enable certification to be carried out correctly. If, however, a voltage operated device is found while preparing a periodic inspection report, a recommendation that it should be replaced would be the correct way of dealing with it.

### BS 4293 General purpose device

These RCDs are very common in installations although they ceased to be used in the early 1990s. They have been replaced by BS EN 61008-1, BS EN 61008-2-1 and BS EN 61008-2-2.

They are used as standalone devices or main switches fitted in consumers' units/distribution boards.

This type of device provides protection against earth fault current. They will commonly be found on TT systems 15 or more years old, although they may be found on TNS systems where greater protection was required.

A typical voltage – operated earth leakage current breaker (ELCB) – now obsolete

A typical general purpose RCD to BS4293 – now obsolete

The problem with using a low tripping current device as the main switch is that nuisance tripping could occur. This type of protection would not be acceptable as compliance with the 17th edition of BS7671 Wiring Regulations. If major alterations were being carried out, then the protection would need to be changed to comply with the modern way of thinking which is explained later in this chapter.

A typical 2-phase
RCCB to BS EN 61008-1

### BS 4293 Type S

These are time delayed RCDs and are used to give good discrimination with other RCDs.

### BS EN 61008-1 General purpose device

This is the current standard for a residual current circuit breaker (RCCB) and provides protection against earth fault current. These devices are generally used as main switches in consumers' units/distribution boards.

Three-phase devices are also very common.

### BS 7288

This is the current standard for RCD-protected socket outlets and provides protection against earth fault currents. These socket outlets would be used in areas where there is an increased risk of electric shock, such as common areas of schools and colleges. It is also a requirement that any socket outlet used for portable equipment outdoors must have supplementary protection provided by an RCD. Where the socket outlets are sited outside, waterproof BS 7288 outlets are used to IP 56.

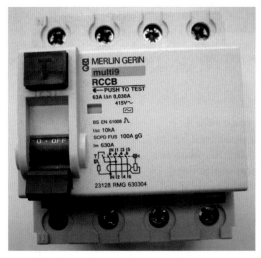

A typical 3-phase RCCB to BS EN 61008-1

## BS EN 61009-1

This is the standard for a residual current circuit breaker with overload (RCCBO) protection.

These devices are generally used to provide single circuits with earth fault protection, overload protection and short circuit protection. They are fitted in place of miniature circuit breakers and the correct type should be used (types B, C or D).

## BS EN 61008-1 Type S

These are time delayed RCDs and are used to give good discrimination with other RCDs.

Section 3 of the *On-Site Guide* gives good examples of how these devices should be used within an installation.

### RCDs and supply systems

#### TT System

BS 7671 states that care must be taken to ensure that the operation of a single protective device should not cause a dangerous situation (Regulation 314.2). One RCD protecting the whole installation is now no longer acceptable in the majority of installations.

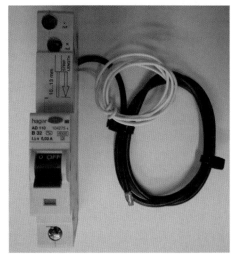

A typical RCCBO to BS EN 61009-1

Compliance with Regulation 314 can be achieved by using a split board with a non-RCD main switch and RCDs protecting both sides of the split board. In this instance, careful consideration should be given to how the circuits are divided, possibly mixing upstairs and downstairs circuits to each side of the board. This would avoid the whole of the upstairs or downstairs circuits having loss of supply due to a fault on a single circuit.

Another method would be to use a consumer unit with a main switch to BSEN 60947-3 and RCBOs to BS EN 610091 as protective devices for all circuits. This option is perfectly satisfactory but can work out a little expensive!

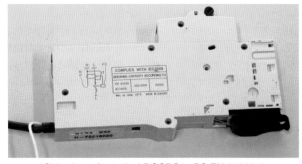

Side view of a typical RCCBC to BS EN 61009-1

## TNS and TNCS systems

The previous options would be suitable for these systems but where RCD protection is not required for all circuits, a standard board with a non-RCD protected main switch could be used. RCCBOs would then be fitted for the circuits requiring RCD protection and normal protective devices used for any circuits not requiring RCD protection.

Other options might be available as different products are introduced by manufacturers of electrical equipment.

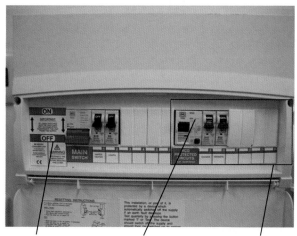

100 mA S-type          30 mA RCD          Only this part is
RCD (BS EN 61008)      (BS EN 61008)      RCD – protected
as main switch         protection

## Testing of RCDs

**Remember that these are live tests and care should be taken whilst carrying them out**

The instrument to be used to carry out this test is an RCD tester, with leads to comply with GS 38.

### Voltage operated (ELCBs)

No test required as they should now be replaced.

### BS 4293 RCDs

If this type of RCD is found on TT systems or other systems where there is a high value of earth fault impedance ($Z_e$), the RCD tester should be plugged into the nearest socket or connected as close as possible to the RCD. The tester should then be set at the rated tripping current of the RCD (I$\Delta$n); for example, at 30 mA (*be careful and do not mistake the tripping current for the current rating of the device*).

Before carrying out the tests ensure that all loads are removed; failure to do this may result in the readings being inaccurate.

## STEP 1

The test instrument must then be set at 50% of the tripping current (15 mA).

## STEP 2

Push the test button of the instrument, the RCD should not trip.

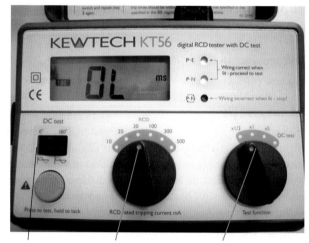

Switch on 180° — Set to rated tripping current of the RCD — Set at 50% of the tripping current

## STEP 3

The test instrument will have a switch on it which will enable the instrument to test the other side of the waveform 0° ~ 80°. This switch must be moved to the opposite side and the test repeated. Again the RCD should not trip.

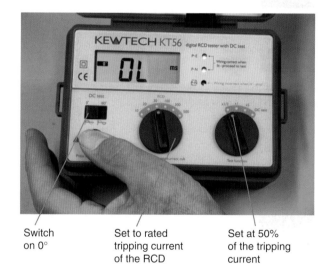

Switch on 0° — Set to rated tripping current of the RCD — Set at 50% of the tripping current

If, while testing an RCD it trips during the 50% test, do not automatically assume that the RCD is at fault.

Consider the possibility that there is a small earth leakage on the circuit or system. Switch all circuits off and test RCD on the load side at 50% using fly leads. If it still trips, then the RCD should be replaced.

If the RDC does not trip, then turn each circuit on one at a time, carrying out a 50% test each time a circuit has been turned on. When the RCD does trip, switch off all circuits except the last one which was switched on. Test again. If the RCD trips carry out an insulation test on this circuit as it probably has a low insulation resistance. If, however, the RCD does not trip it could be an accumulation of earth leakage from several circuits and they should all be tested for insulation resistance.

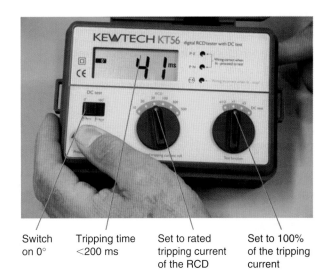

### STEP 4

Now set the test current to the rated tripping current (30 mA).

Switch on 0°    Tripping time <200 ms    Set to rated tripping current of the RCD    Set to 100% of the tripping current

### STEP 5

Push the test button, and the RCD should trip within 200 milliseconds.

### STEP 6

Reset the RCD and the slowest time in which it tripped should be entered on to the test result schedule.

### STEP 7

Set the test current to five times the rated tripping current (150 mA).

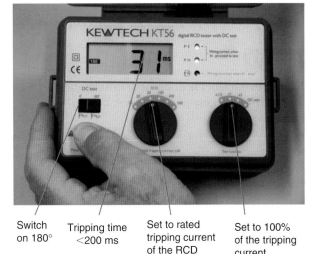

### STEP 8

Push the test button and the RCD should trip within 40 milliseconds.

Switch on 180°    Tripping time <200 ms    Set to rated tripping current of the RCD    Set to 100% of the tripping current

### STEP 9

Move the waveform switch to the opposite side, and repeat the test. Again it must trip within 40 milliseconds (five times faster than the times 1 test).

### STEP 10

Push the test button and the RCD should trip within 40 milliseconds.

### STEP 11

Move the waveform switch to the opposite side, and repeat the test. Again it must trip within 40 milliseconds (five times faster than the times 1 test).

**After completion of the instrument tests**

### STEP 12

Push integral test button on RCD to verify that the mechanical parts are working correctly.

### STEP 13

Ensure that a label is in place to inform the user of the necessity to use the test button quarterly.

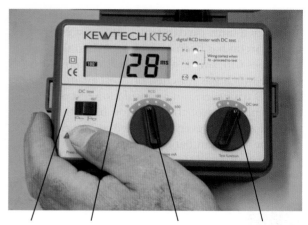

Switch on 180°    Tripping time <40ms    Set to rated tripping current of the RCD    Set to 5 times the rated tripping current

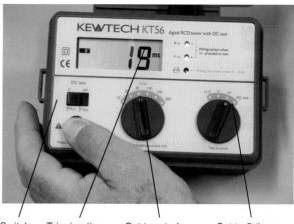

Switch on 0°    Tripping time <40 m    Set to rated tripping current of the RCD    Set to 5 times the rated tripping current

### BS EN 610081

These devices should be tested in exactly the same manner as BS 4293 using the same test instrument. However, the difference is that, when carrying out the 100% test, the tripping time is increased to 300 milliseconds.

### BS 4293 Type S

This device has a built in time delay. The simple way to think about this is that it does not recognize a fault for 200 milliseconds, and they must trip within 200 milliseconds after that.

**STEP 1**   Plug in or connect an RCD as close as possible to the RCD to be tested.

**STEP 2**   Set the instrument on the trip current of the RCD and ensure that it is set for 'S' type.

**STEP 3**   Test at 50% and the device should not trip.

**STEP 4**   Repeat the test on the opposite waveform.

**STEP 5**   Set test instrument on 100% and carry out test. The RCD should trip within 400 milliseconds (200 *ms time delay and 200 ms fault*).

**STEP 6**   Repeat on the opposite wave form.

The slowest operating time at the 100% test should be recorded, as should the fact that it is an 'S' type.

### BS EN 61008 Type S

This device has a time delay of 200 milliseconds and a tripping time of 300 milliseconds, making a maximum tripping time of 500 milliseconds.

The test should be carried out as the BS 4293 Type S but remember the different tripping time.

### BS 7288 RCD protected socket

This device should be tested the same as for a BS 4293 and the tripping times are the same.

> The 5 times test must only be carried out on RCDs with trip ratings (IΔn) up to 30 mA.

> Always ensure that it is safe to carry out these tests. Remember to remove any loads, ensure that the disconnecting of the supply due to the test will not effect any equipment or cause damage. If any people are within the building ensure that they are aware of testing being carried out, and that a loss of supply is likely.

Consideration should be given to whether the socket will supply portable equipment outdoors. If it can it should be tested at 5 times its rating.

### *BS EN 61009 RCBOs*

These devices should be tested as for BS 4239 RCDs but the disconnection times are:

- 50% test on both sides of waveform, no trip.
- 100% test on both sides of waveform; must trip within 300 milliseconds.
- If used as supplementary protection, the 5 times test must also be carried out; it must trip within 40 milliseconds.

The product standard performance criteria can be found in Table 3A in Appendix 3 of BS 7671.

# Completion of test certificates

The following pages detail the test certificates and itemized descriptions. At the end of the chapter further 'Notes for Recipients' can be found.

## MINOR DOMESTIC ELECTRICAL INSTALLATION WORKS CERTIFICATE

This certificate is to be completed when additions to existing circuits have been carried out. For example, an additional lighting point or socket outlet.

If more than one circuit has been added to, then a separate Minor Works Certificate must be issued for each modification.

These certificates vary slightly depending on which certification body has supplied them; some require slightly more information than others. The information required is as follows.

**DOMESTIC INSTALLER**

This safety certificate is an important and valuable document which should be retained for future reference

This certificate is not valid if the serial number has been defaced or altered

**DMP3/**

# MINOR DOMESTIC ELECTRICAL INSTALLATION WORKS CERTIFICATE

Issued in accordance with *British Standard 7671 – Requirements for Electrical Installations* by an Approved Contractor or Conforming Body enrolled with NICEIC, Warwick House, Houghton Hall Park, Houghton Regis, Dunstable, LU5 5ZX.

**To be used only for minor electrical work which does not include the provision of a new circuit**

## PART 1: DETAILS OF THE MINOR WORKS

Client

Date minor works completed

Contract reference, if any

Description of the minor works

Details of departures, if any, from BS 7671 (as amended)

Location/address of the minor works

Postcode

## PART 2: DETAILS OF THE MODIFIED CIRCUIT

| | | | |
|---|---|---|---|
| System type and earthing arrangements | TN-C-S | TN-S | TT |
| Protective measures against electric shock | ADS | Other | |
| Overcurrent protective device for the modified circuit | BS(EN) | Type | Rating   A |
| Residual current device (if applicable) | BS(EN) | Type | $I_{\Delta n}$   mA |

Details of wiring system used to modify the circuit   Type    Reference method    csa of live conductors   $mm^2$   csa of cpc   $mm^2$

Where the protective measure against electric shock is ADS, insert maximum disconnection time permitted by BS 7671   s   Maximum $Z_s$ permitted by BS 7671   $\Omega$

Comments, if any, on existing installation

## PART 3: INSPECTION AND TESTING OF THE MODIFIED CIRCUIT AND RELATED PARTS   †*Essential inspections and tests*

| | | | |
|---|---|---|---|
| † Confirmation that necessary inspections have been undertaken | (✓) | † Confirmation of the adequacy of earthing | (✓) |
| † Circuit resistance   $R_1 + R_2$   $\Omega$   or   $R_2$   $\Omega$ | | † Confirmation of the adequacy of protective bonding | (✓) |
| Insulation resistance (* *In a multi-phase circuit, record the lower or lowest value, as appropriate*)   Line/Line* | $M\Omega$ | † Confirmation of correct polarity | (✓) |
|   Line/Neutral† | $M\Omega$ | † Maximum measured earth fault loop impedance, $Z_s$ | $\Omega$ |
| Instrument Serial No(s)   † Line/Earth* | $M\Omega$ | † RCD operating time at $I_{\Delta n}$ (if RCD fitted) | ms |
|   † Neutral/Earth† | $M\Omega$ | RCD operating time at 5 x $I_{\Delta n}$, if applicable | ms |

Agreed limitations, if any, on the inspection and testing

## PART 4: DECLARATION

I/We CERTIFY that the said works do not impair the safety of the existing installation, that the said works have been designed, constructed, inspected and tested in accordance with BS 7671:   (IEE Wiring Regulations), amended to   and that the said works, to the best of my/our knowledge and belief, at the time of my/our inspection complied with BS 7671 except as detailed in Part 1.

The results of the inspection and testing reviewed by the Qualified Supervisor

For and on behalf of *(Trading Title of Domestic Installer)*

| Name (CAPITALS) | | Name (CAPITALS) | | Address |
|---|---|---|---|---|
| Signature | | Signature | | |
| Date | | Date | | Postcode |
| Registration Number | | | | *(The registration number is essential information)* |

This form is based on the model shown in Appendix 6 of BS 7671 (as amended)
Published by NICEIC Group Limited © Copyright The Electrical Safety Council (Jan 2008)

**Please see the 'Notes for Recipients' on the reverse of this page.**

DMP3/1

**Client** Name of the person ordering the work.

**Location/address** The address at which the work is carried out.

**Date** Completion of minor work.

**Description** It is important to document exactly the work which has been carried out.

**Type of system** TT, TNS, TNCS.

**Method of protection against indirect contact** This will usually be EEBAADS.

**Overcurrent device for the circuit** This is the type and size of device which is protecting the circuit on which the minor works has been carried out. If it is necessary to change the protective device, then an electrical installation certificate is required. Not a Minor Works Certificate.

**Residual current device** Type, current rating and tripping value $I\Delta n$ is required. If the addition to the circuit requires the fitting of an RCD, then an Electrical Installation Certificate is required for the RCD.

**Details of wiring system used** What type of wiring is it? For example, PVC, conduit, steel wire armour.

**Reference method** How has it been installed? See Appendix 4, BS 7671 for reference methods.

**CSA of conductors** What size are the conductors?

**Maximum disconnection time of the circuit** Is it 0.2, 0.4, 1 or 5 seconds?

**Maximum $Z_S$** What is the maximum $Z_S$ permitted to ensure that the protective device operates in the correct time? These can be found in Part 4 of BS 7671.

**Circuit resistance** What is the value of $R_1 + R_2$ or $R_2$ if Table 41C of BS 7671 is being used.

**Confirmation of bonding** Has bonding been installed? If not, this should be pointed out to the client, and identified in the comments section of the certificate. *It is still permissible to carry out minor works if bonding is not present in the installation.*

**Confirmation of earthing** Is the installation earthed? If not, then the work should not be carried out and the client should be informed of the danger that this presents.

**Correct polarity** Is the supply correct? Have the circuit conductors been connected in to the correct terminals?

**Measured $Z_S$** What is the measured value of $Z_S$ for the altered circuit? Check that it is lower than the maximum permitted value.

**RCD operating time at I$\Delta$n** Must be less than 300 ms if BS EN type, or 200 ms if BS type.

**RCD operating time at 5 I$\Delta$n** Only required for RCDs rated at 30 mA or less, when used for supplementary protection against direct contact. Never required on RCDs above 30 mA.

**Insulation resistance** Values required for between live conductors and live conductors and earth. If the measured value is below $2\,\text{M}\Omega$, further investigation is required. For 3 phase circuits record the lowest value.

**Comments on existing installation** Generally, just a comment on the visual condition of the installation; such as, is it old? Perhaps a periodic inspection report may be advised. Is the earthing up to the current requirements necessary for BS 7671?

**Agreed limitations on the inspecting and testing** Not usually many on a minor works. Could possibly be where it is difficult to disconnect or isolate vulnerable equipment and the insulation resistance test is carried out between live conductors joined and earth only.

# NOTES FOR RECIPIENTS

TRACEABLE
SERIAL NUMBER

## THIS SAFETY CERTIFICATE IS AN IMPORTANT AND VALUABLE DOCUMENT
## WHICH SHOULD BE RETAINED FOR FUTURE REFERENCE

IF YOU WERE THE PERSON ORDERING THE WORK, BUT NOT THE OWNER OR USER OF THE INSTALLATION, YOU SHOULD PASS THIS CERTIFICATE, OR A FULL COPY OF IT INCLUDING THESE NOTES, IMMEDIATELY TO THE OWNER OR USER OF THE INSTALLATION.

This safety certificate has been issued to confirm that the minor electrical installation works to which it relates has been designed, constructed, inspected, tested and verified in accordance with the national standard for the safety of electrical installations, British Standard 7671 (as amended) - *Requirements for Electrical Installations* (the IEE Wiring Regulations).

Where, as will often be the case, the existing installation incorporates a residual current device (RCD), there should be a notice at or near the main switchboard or consumer unit stating that the device should be tested at quarterly intervals. For safety reasons, it is important that you carry out the test regularly.

Also, for safety reasons, the complete electrical installation including the minor electrical installation works which is the subject of this certificate will need to be inspected and tested at appropriate intervals by a competent person. NICEIC* recommends that you engage the services of an NICEIC Approved Contractor for this purpose. There should be a notice at or near the origin of the existing installation (such as at the consumer unit or main switchboard) which indicates when the inspection of the complete installation is next due.

Only the NICEIC Domestic Installer responsible for the work is authorised to issue this NICEIC certificate. The certificate has a printed serial number which is traceable to the Domestic Installer to which it was supplied by NICEIC.

You should have received the certificate marked 'Original' and the Domestic Installer should have retained the certificate marked 'Duplicate'. **The 'Original' certificate should be retained in a safe place and shown to any person inspecting, or undertaking further work on, the electrical installation in the future. If you later vacate the property, this certificate will demonstrate to the new user that the minor electrical installation works complied with the requirements of the national electrical safety standard at the time the certificate was issued.**

The Minor Domestic Electrical Installation Works Certificate is intended to be used only for an addition or alteration to an existing circuit that does not extend to the provision of a new circuit. Examples include the addition of a socket-outlet to an existing circuit or the addition of a lighting point to an existing circuit, or the replacement or relocation of a light switch. A separate certificate should have been received for each existing circuit on which minor works has been carried out. This certificate would be considered by NICEIC to be invalid if you requested the Domestic Installer to undertake more extensive work, for which a Domestic Electrical Installation Certificate should have been issued.

Part 3 of this certificate is intended to facilitate the recording of information associated with the inspection and testing of the modified circuit, and the related parts of the existing installation on which the modified circuit depends for its safety. Generally, each box should have been completed to confirm the results of a particular inspection or test by a 'Yes' or a '✓', or by the insertion of a measured value. Where a particular inspection or test was not applicable, this should have been indicated by 'N/A', meaning 'Not Applicable'. Where an inspection or a test was not practicable, the entry should read 'LIM', meaning 'Limitation', acknowledging that the particular circumstances prevented the particular inspection or test procedure from being carried out. In such a case, each limitation should have been recorded in the box entitled 'Agreed limitations, if any, on the inspection and testing', together with the reason for each limitation.

If wiring alterations or additions are made to an installation such that wiring colours to two versions of BS 7671 exist, a warning notice should have been affixed at or near the appropriate consumer unit.

Should the person ordering the work (eg the client, as identified on this certificate), have reason to believe that any element of the work for which the Domestic Installer has accepted responsibility (as indicated by the signature on this certificate) does not comply with the requirements of the national electrical safety standard (BS 7671), the client should in the first instance raise the specific concerns in writing with the Domestic Installer. If the concerns remain unresolved, the client may make a formal complaint to NICEIC, for which purpose a standard complaint form is available on request.

The complaints procedure offered by NICEIC is subject to certain terms and conditions, full details of which are available upon application and from the website[†]. NICEIC does not investigate complaints relating to the operational performance of electrical installations (such as lighting levels), or to contractual or commercial issues (such as time or cost).

*NICEIC' is a trading name of NICEIC Group Limited, a wholly owned subsidiary of The Electrical Safety Council. Under licence from The Electrical Safety Council, NICEIC acts as the electrical contracting industry's independent voluntary regulatory body for electrical installation safety matters throughout the UK, and maintains and publishes registers of electrical contractors that it has assessed against particular scheme requirements (including the technical standard of electrical work).*

---

### For further information about electrical safety and how the NICEIC can help you, visit **www.niceic.com**

---

DMP2/1&2B

> This certification is required for a new installation or circuit

## ELECTRICAL INSTALLATION CERTIFICATE

This certificate is to be completed for a new circuit, a new installation, a rewiring and any circuit where the protective device has been changed.

In the case of a consumer's unit change only, an Electrical Installation Certificate would be required for the consumers unit and a Periodic Inspection Report should be completed for the existing installation.

A standard Electrical Installation Certificate can be used for any installation. However, if the work to be certificated is covered by building regulation, a Part P Certificate is available solely for this purpose. These certificates simplify the paper work by including a schedule of inspection and a Schedule of Test Results on the same document.

A Schedule of Test Results and a Schedule of Inspection must be completed to accompany an electrical installation certificate and a periodic inspection report.

These certificates vary slightly depending on which certification body has supplied them, some require slightly more information than others. The Electrical Installation Certificate and Particulars of Signatures to the Electrical Installation Certificate are typical of an electrical installation certificate. The information required is as follows.

**ICM4**

# ELECTRICAL INSTALLATION CERTIFICATE

Issued in accordance with *British Standard BS 7671- Requirements for Electrical Installations*

**This safety certificate is an important and valuable document which should be retained for future reference**

## DETAILS OF THE CLIENT

Client / Address:

## DETAILS OF THE INSTALLATION

The installation is:

Address:

Extent of the installation covered by this certificate:

New

An addition

An alteration

## DESIGN

I/We, being the person(s) responsible for the design of the electrical installation (as indicated by my/our signature(s) below), particulars of which are described above, having exercised reasonable skill and care when carrying out the design, hereby CERTIFY that the design work for which I/we have been responsible is to the best of my/our knowledge and belief in accordance with BS 7671: amended to (date) except for the departures, if any, detailed as follows:

Details of departures from BS 7671, as amended (Regulations 120.3, 120.4):

The extent of liability of the signatory/signatories is limited to the work described above as the subject of this certificate.
For the **DESIGN** of the installation:                    **\*\*(Where there is divided responsibility for the design)**

| Signature | Date | Name (CAPITALS) | Designer 1 |
| Signature | Date | Name (CAPITALS) | \*\* Designer 2 |

## CONSTRUCTION

I/We, being the person(s) responsible for the construction of the electrical installation (as indicated by my/our signature below), particulars of which are described above, having exercised reasonable skill and care when carrying out the construction, hereby CERTIFY that the construction work for which I/we have been responsible is to the best of my/our knowledge and belief in accordance with BS 7671: amended to (date) except for the the departures, if any, detailed as follows:

Details of departures from BS 7671, as amended (Regulations 120.3,120.4):

The extent of liability of the signatory is limited to the work described above as the subject of this certificate.
For the **CONSTRUCTION** of the installation:

| Signature | Date | Name (CAPITALS) | Constructor |

## INSPECTION AND TESTING

I/We, being the person(s) responsible for the inspection and testing of the electrical installation (as indicated by my/our signatures below), particulars of which are described above, having exercised reasonable skill and care when carrying out the inspection and testing, hereby CERTIFY that the work for which I/we have been responsible is to the best of my/our knowledge and belief in accordance with BS 7671: amended to (date) except for the departures, if any, detailed as follows:

Details of departures from BS 7671, as amended (Regulations 120.3,120.4):

The extent of liability of the signatory/signatories is limited to the work described above as the subject of this certificate.
For the **INSPECTION AND TESTING** of the installation:          Reviewed by †

| Signature | Date | Signature | Date |
| Name (CAPITALS) | Inspector | Name (CAPITALS) | |

## DESIGN, CONSTRUCTION, INSPECTION AND TESTING \*

\* This box to be completed only where the design, construction, inspection and testing have been the responsibility of one person.

I, being the person responsible for the design, construction, inspection and testing of the electrical installation (as indicated by my signature below), particulars of which are described above, having exercised reasonable skill and care when carrying out the design, construction, inspection and testing, hereby CERTIFY that the said work for which I have been responsible is to the best of my knowledge and belief in accordance with BS 7671, amended to (date) except for the departures, if any, detailed as follows:

Details of departures from BS 7671, as amended (Regulations 120.3, 120.4):

The extent of liability of the signatory is limited to the work described above as the subject of this certificate.
For the **DESIGN**, the **CONSTRUCTION** and the **INSPECTION AND TESTING** of the installation.          Reviewed by †

| Signature | Date | Signature | Date |
| Name (CAPITALS) | | Name (CAPITALS) | |

† The completed schedules of inspection and testing should preferably be reviewed by another competent person to confirm that the recorded results are consistent with electrical installation work conforming to the requirements of BS 7671

Page 1 of [   ]

**Please see the 'Notes for Recipients' on the reverse of this page.**

ICM4/1

**Details of client** Name and address of the person ordering the work.

**Location/address** The address at which the work is carried out.

**Details of the installation** What part of the installation does this certificate cover: is it all of the installation, or is it a single circuit? It is vital that this part of the certificate is completed as accurately as possible.

There are generally three tick boxes regarding the nature of the installation.

**New** To be ticked if the whole installation is new. This would include a rewire.

**Alteration** To be ticked where the characteristics of an existing circuit have been altered (such as extending/altering a circuit and changing the protective device). This box would also cover the replacement of consumer units and/or the fitting of RCDs.

**Addition** Used to identify when a new circuit or numerous circuits have been added to an existing circuit.

**Design, construction, inspection and testing** The person or persons responsible for each of these must sign. It could be one person or possibly three, depending on the job. However it is important that all boxes have a signature.

Usually in this section there will be two boxes referring to BS 7671. To complete this correctly, look at the top right-hand corner of BS 7671 – where the words 'BS 7671 "Year"' will be seen. Just below this will be the date of the amendments. This will indicate the most recent amendment.

**Next inspection** The person who has designed the installation, or the part of it that this certificate covers must recommend when the first periodic inspection and test is carried out. This will be based on the type of use to which it will be put, and the type of environment.

# SCHEDULE OF CIRCUIT DETAILS
# FOR THE INSTALLATION

| TO BE COMPLETED IN EVERY CASE | TO BE COMPLETED ONLY IF THE DISTRIBUTION BOARD IS NOT CONNECTED DIRECTLY TO THE ORIGIN OF THE INSTALLATION* |
|---|---|

| | |
|---|---|
| Location of distribution board: | Supply to distribution board is from: |
| | Overcurrent protective device for the distribution circuit: |
| Distribution board designation: | Type: BS(EN) |

No of phases:

Nominal voltage: V

Associated RCD (if any): BS(EN)

Rating: A

RCD No of poles:

$I_{\Delta n}$ mA

## CIRCUIT DETAILS

| Circuit number and phase | Circuit designation | Type of wiring (see code below) | Reference method ↑ | Number of points served | Circuit conductors: csa | | Max. disconnection time permitted by BS 7671 (s) | Overcurrent protective devices | | | | | | Maximum $Z_s$ permitted by BS 7671 (Ω) |
|---|---|---|---|---|---|---|---|---|---|---|---|---|---|---|
| | | | | | Live (mm²) | cpc (mm²) | | BS (EN) | Type No | Rating (A) | Short-circuit capacity (kA) | RCD Operating current $I_{\Delta n}$ (mA) | | |
| | | | | | | | | | | | | | | |
| | | | | | | | | | | | | | | |
| | | | | | | | | | | | | | | |
| | | | | | | | | | | | | | | |
| | | | | | | | | | | | | | | |
| | | | | | | | | | | | | | | |
| | | | | | | | | | | | | | | |
| | | | | | | | | | | | | | | |
| | | | | | | | | | | | | | | |
| | | | | | | | | | | | | | | |
| | | | | | | | | | | | | | | |
| | | | | | | | | | | | | | | |
| | | | | | | | | | | | | | | |
| | | | | | | | | | | | | | | |
| | | | | | | | | | | | | | | |
| | | | | | | | | | | | | | | |
| | | | | | | | | | | | | | | |
| | | | | | | | | | | | | | | |
| | | | | | | | | | | | | | | |
| | | | | | | | | | | | | | | |
| | | | | | | | | | | | | | | |

↑ *See Table 4A2 of Appendix 4 of BS 7671: 2008*

| CODES FOR TYPE OF WIRING | | | | | | | | |
|---|---|---|---|---|---|---|---|---|
| A | B | C | D | E | F | G | H | O (Other - please state) |
| PVC/PVC cables | PVC cables in metallic conduit | PVC cables in non-metallic conduit | PVC cables in metallic trunking | PVC cables in non-metallic trunking | PVC/SWA cables | XLPE/SWA cables | Mineral-insulated cables | |

Page 4 of ☐

* *In such cases, details of the distribution (sub-main) circuit(s), together with the test results for the circuit(s), must also be provided on continuation schedules.*

This form is based on the model shown in Appendix 6 of BS 7671: 2008.
© Copyright The Electrical Safety Council (Jan 2008).

See next page for
Schedule of Test Results

ICM4/7

**Location of distribution board** Where is it?

**Designation of consumer unit** If there is more than one unit how is it identified – number, name or letter?

**Circuit designation** What does the circuit supply? Is it a cooker, ring, etc?

If the circuit is fed by a submain (*distribution circuit*), details of the sub-main must be recorded. Possibly on a separate schedule, or on the top line of the schedule which you are completing.

**Type of wiring** Is it PVC twin and CPC, plastic or steel conduit? Some certification bodies have their own codes for this.

**Reference method** How is it installed? The methods are detailed in Appendix 4 of BS 7671.

**Number of points served** How many outlets or items of fixed equipment are on the circuit?

**Circuit conductor size** What size are live conductors and CPC? (give in $mm^2$).

## Maximum disconnection time permissible for the circuit

Refer to Table 41.1 of BS7671.

### TT systems
Final circuits that are rated up to 32A require a maximum disconnection time of 0.2 seconds. Distribution circuits require a maximum disconnection time of 1 second.

### TN systems
Final circuits that are rated up to 32A require a maximum disconnection time of 0.4 seconds. Distribution circuits require a maximum disconnection time of 5 seconds.

### Type of over-current protective device
BS or BS EN. Enter the number.

**Type** If BS 1361 it will be a number 1, 2, 3 or 4; if BS EN it will be B, C, or D. The letter A was not used in case it was mistaken for Amperes. If the device is a fuse '/' should be entered.

**Rating** What is the current rating of the device?

**Short circuit capacity** What is the short circuit capacity of the device? It may be marked on the device; if it is not, then Table 7.4 in the *On-Site Guide* will be of assistance.

Whatever it is, it must be at least equal to the PFC measured at the main switch of the consumer unit.

## Maximum $Z_S$

This is the maximum $Z_S$ permitted by BS 7671 for the protective device in this circuit. This value can be found in Tables 41.2, 41.3 or 41.4 in the *On-Site Guide*. Be careful to use the correct table relating to the disconnection time for the circuit if fuses are used.

# SCHEDULE OF TEST RESULTS
# FOR THE INSTALLATION

<table>
<tr><td colspan="3">TO BE COMPLETED ONLY IF THE DISTRIBUTION BOARD IS NOT CONNECTED DIRECTLY TO THE ORIGIN OF THE INSTALLATION</td><td colspan="2">Test instruments (serial numbers) used:</td></tr>
<tr><td colspan="3">Characteristics at this distribution board</td><td></td><td></td></tr>
<tr><td colspan="3">Confirmation of supply polarity</td><td>Earth fault loop impedance</td><td>RCD</td></tr>
<tr><td>★ See note below<br>$Z_s$ *</td><td>Ω</td><td>Operating times of associated RCD (if any)   At $I_{\Delta n}$   ms</td><td>Insulation resistance</td><td>Other</td></tr>
<tr><td>$I_{pf}$ *</td><td>kA</td><td>At 5I$_{\Delta n}$ (if applicable)   ms</td><td>Continuity</td><td>Other</td></tr>
</table>

## TEST RESULTS

| Circuit number and phase | Circuit impedances (Ω) | | | | | Insulation resistance † *Record lower or lowest value* | | | | Polarity | Maximum measured earth fault loop impedance, $Z_s$ ★ *See note below* | RCD operating times | |
|---|---|---|---|---|---|---|---|---|---|---|---|---|---|
| | Ring final circuits only (measured end to end) | | | All circuits (At least one column to be completed) | | Line/Line † | Line/Neutral † | Line/Earth † | Neutral/Earth | | | at $I_{\Delta n}$ | at 5I$_{\Delta n}$ (if applicable) |
| | $r_1$ (Line) | $r_n$ (Neutral) | $r_2$ (cpc) | $R_1 + R_2$ | $R_2$ | (MΩ) | (MΩ) | (MΩ) | (MΩ) | (✓) | (Ω) | (ms) | (ms) |
| | | | | | | | | | | | | | |
| | | | | | | | | | | | | | |
| | | | | | | | | | | | | | |
| | | | | | | | | | | | | | |
| | | | | | | | | | | | | | |
| | | | | | | | | | | | | | |
| | | | | | | | | | | | | | |
| | | | | | | | | | | | | | |
| | | | | | | | | | | | | | |
| | | | | | | | | | | | | | |
| | | | | | | | | | | | | | |
| | | | | | | | | | | | | | |
| | | | | | | | | | | | | | |
| | | | | | | | | | | | | | |
| | | | | | | | | | | | | | |
| | | | | | | | | | | | | | |
| | | | | | | | | | | | | | |
| | | | | | | | | | | | | | |
| | | | | | | | | | | | | | |
| | | | | | | | | | | | | | |
| | | | | | | | | | | | | | |
| | | | | | | | | | | | | | |

★ Note: Where the installation can be supplied by more than one source, such as a primary source (eg public supply) and a secondary source (eg standby generator), the higher or highest values must be recorded.

**TESTED BY**

| Signature: | Position: | Page 5 of ☐ |
|---|---|---|
| Name: (CAPITALS) | Date of testing: | |

See previous page for Circuit Details

ICM4/9

## Circuit impedances

If the circuit is a ring final circuit then $R_1$, $r_n$ and $R_2$ must be recorded on some certificates where boxes are provided. On other certificates, $R_1 + R_2$ only may be recorded. This is the measured end to end value of the respective conductor. If the circuit is not a ring final circuit then enter 'n/a' or '/'.

$R_1 + R_2$ This value must be entered for all circuits unless it is not possible to measure. Usually this value is obtained when carrying out the continuity of CPC test.

$R_2$ Where the measurement of $R_1 + R_2$ is not possible then the end to end resistance of the CPC can be measured using the long lead method.

## Insulation resistance

This is the value of insulation resistance in $m\Omega$ measured between the conductors as identified in the heading. If the installation is measured from the tails, providing the value is greater than the range of the instrument ($>200\,m\Omega$ for example). Then this value can be used for all of the circuits. If the value is less than the range of the instrument, then it would be better to split the installation and measure each circuit individually. A value must be entered; infinity readings are not valid.

Line to Line readings are for 3 phase circuits.

## Polarity

This box is just a tick box to confirm that you have checked polarity. This is normally done when carrying out the continuity of CPC tests.

Live polarity of the incoming supply should be tested at the main switch.

## Measured earth fault loop impedance $Z_S$

This is the measured value of the circuit. The measurement should be taken at the furthest point of the circuit.

It should be compared with the $Z_e + R_1 + R_2$ total and if it is higher, then further investigation should be carried out.

The measured value of $Z_e + R_1 + R_2$ will not include parallel paths if carried out correctly; whereas the measured $Z_S$ will, as this is a live test and all protective conductors must be connected for the test to be carried out safely. Therefore, the measured $Z_S$ should be the same as the $Z_e + R_1 + R_2$ value or even less if parallel paths are present. It should not be higher!

### RCD operating times

At IΔn: the actual operating time at the trip rating should be entered here.

At 5 IΔn: this value is only applicable for RCDs with a trip rating of up to 30 mA and should not be carried out on RCDs with a higher rating.

### Other

Under the heading of RCD operating times, there is often a column labelled 'other'. This is where the correct mechanical operation of switches, circuit breakers and isolators, etc. are recorded.

### Remarks

This area is for the inspector to record anything about the circuit that he/she feels necessary. It may not be a fault but possibly something that may be useful to the next person carrying out an inspection and test on the installation.

#### SCHEDULE OF ITEMS INSPECTED

This certificate, along with the Schedule of Test Results, forms part of the Electrical Installation Certificate and the Periodic Test Report. Without these schedules the other certificates and reports are invalid.

Completion of this certificate involves the completion of boxes which must be marked using a '✓', 'X' or 'n/a' after the inspection is made, and could be useful as a checklist.

An 'X' should never be entered on to a schedule which accompanies an IEC.

## F. OBSERVATIONS AND RECOMMENDATIONS FOR ACTIONS TO BE TAKEN

**Referring to the attached schedules of inspection and test results, and subject to the limitations at D:**

There are no items adversely affecting electrical safety.

**or**

The following observations and recommendations are made.

| Item No | | Code † |
|---------|---|--------|
| **1** | | |

*Note: If necessary, continue on additional pages(s), which must be identified by the Periodic Inspection Report date and page number(s).*

† *Where observations are made, the inspector will have entered one of the following codes against each observation to indicate the action (if any) recommended:-*

1. *'requires urgent attention' or*
2. *'requires improvement' or*
3. *'requires further investigation' or*
4. *'does not comply with BS 7671'*

**Please see the reverse of this page for guidance regarding the recommendations.**

**Urgent remedial work recommended for Items:** | **Corrective action(s) recommended for Items:**

## G. SUMMARY OF THE INSPECTION

General condition of the installation:

*Note: If necessary, continue on additional page(s), which must be identified by the Periodic Inspection Report date and page number(s).*

Date(s) of the inspection: | Overall assessment of the installation:

*(Entry should read either 'Satisfactory' or 'Unsatisfactory')*

Page 2 of

**Please see the 'Guidance for Recipients on the Recommendation Codes' on the reverse of this page.**

IPM3/3

*Observations and recommendations*

Defects, if any, must be recorded accurately here and a code given to them. Some certificates have codes to indicate the level of attention required. On most certificates the codes will be:

1. **Requires urgent attention** Anything which could compromise the safety of those using the installation should be entered here. This would include lack of earthing, undersized cables, damaged accessories, high $Z_s$ values. It is up to the person carrying out the inspection to make a judgement on this.

2. **Requires improvement** Defects which do not immediately cause the installation to be regarded as unsafe but which could be problematic in the future. This could be: corrosion; old cables such as lighting cables with no CPC terminated in wooden switch boxes; labels missing, etc. Again a judgement must be made by the person carrying out the inspection.

3. **Requires further investigation** This could be anything that the person who is carrying out the inspection and test is concerned about, but which is outside of the agreed extent of the inspection. Possibly a circuit which cannot be traced, or instrument values within the required parameters – anything that might cause concern.

4. **Does not comply with BS 7671** As the requirements of BS 7671 are amended, parts of the installation which would have complied when the installation was new, may now not comply. Examples of this could be: socket outlets that could be used to supply portable equipment outdoors, not protected by an RCD, CPCs which have been sleeved with green sleeving and not green and yellow. Switch returns not identified on older installations.

It must be remembered that it is the responsibility of the person carrying out the inspection to decide on which code to give, the decision should be made using the inspector's experience and common sense.

A true and accurate reflection of the installation must be recorded here. It may require additional pages to explain in detail any observations and recommendations. Do not be influenced by cost or the difficulty in rectifying any defects. The person signing the certificate will be responsible for its content.

## Summary of the inspection

This section of the certificate is to detail the overall condition of the installation. It is often easier for the inspector to break the installation into specific areas, for instance:

Any change of use or environment which may have had an affect on the installation
Earthing arrangements
Bonding
Isolation
Age
Safety

## Overall assessment

This will either be satisfactory or unsatisfactory. In general terms, if the observation area of the form has any defects other than code 4, the assessment must be unsatisfactory.

## H. SCHEDULES AND ADDITIONAL PAGES

Schedule of Items Inspected and Schedules of Items Tested: Page No 4

Schedule of Circuit Details for the Installation: Page No(s) **5**

Additional pages, including additional source(s) data sheets: Page No(s)

Schedule of Test Results for the Installation: Page No(s) **6**

The pages identified here form an essential part of this report. The report is valid only if accompanied by all the schedules and additional pages identified above.

## I. NEXT INSPECTION

I/We recommend that this installation is further inspected and tested after an interval of not more than

*(Enter interval in terms of years, months or weeks, as appropriate)*

**provided that any items at F which have been attributed a Recommendation Code 1 (*requires urgent attention*) are remedied without delay. Items which have been attributed a Recommendation Code 2 or 3 should be actioned as soon as practicable (see F).**

## J. DETAILS OF ELECTRICAL CONTRACTOR

Trading Title:

Address:

Telephone number:

Fax number:

Postcode:

## K. SUPPLY CHARACTERISTICS AND EARTHING ARRANGEMENTS       *Tick boxes and enter details, as appropriate*

**✧ System Type(s)**

TN-S

TN-C-S

TN-C

TT

IT

**✧ Number and Type of Live Conductors**

a.c.

1-phase (2 wire)

2-phase (3 wire)

3-phase (3 wire)

Other    Please state

d.c.

1-phase (3 wire)

3-phase (4 wire)

2 pole

3-pole

other

**Nature of Supply Parameters**

Nominal voltage(s): $U^{(1)}$     V    $U_o^{(1)}$     V

Nominal frequency, $f^{(1)}$     Hz

Prospective fault current, $I_{pf}^{(2)(3)}$     kA

External earth fault loop impedance, $Z_e^{(3)(4)}$     Ω

Number of supplies

*Notes:*

(1) by enquiry

(2) by enquiry or by measurement

(3) where more than one supply, record the higher or highest values

(4) by measurement

**✧ Characteristics of Primary Supply Overcurrent Protective Device(s)**

BS(EN)

Type

Nominal current rating     A

Short-circuit capacity     kA

## L. PARTICULARS OF INSTALLATION AT THE ORIGIN       *Tick boxes and enter details, as appropriate*

**✧ Means of Earthing**

Supplier's facility:

Installation earth electrode:

**Details of Installation Earth Electrode (where applicable)**

Type: (eg rod(s), tape etc)

Electrode resistance, $R_A$:     (Ω)

Location:

Method of measurement:

**✧ Main Switch or Circuit-Breaker**
* (applicable only where an RCD is suitable and is used as a main circuit-breaker)

Type: BS(EN)

No of Poles

Supply conductors: material

Supply conductors: csa     mm²

Voltage rating     V

Current rating, $I_n$     A

RCD operating current, $I_{\Delta n}$     mA

RCD operating time (at $I_{\Delta n}$) *     ms

**Maximum Demand (Load):**     A per phase

**Method of Protection against Indirect Contact:**

**Main Protective Conductors**

Earthing conductor

Conductor material

Conductor csa     mm²

Continuity check     (✓)

Main equipotential bonding conductors

Conductor material

Conductor csa     mm²

Continuity check     (✓)

Bonding of extraneous-conductive-parts (✓)

Water service          Gas service

Oil service          Structural steel

Lightning protection          Other incoming service(s)

✧ *Where a number of sources are available to supply the installation, and where the data given for the primary source may differ from other sources, a separate sheet must be provided which identifies the relevant information relating to each additional source.*

Page 3 of

This form is based on the model shown in Appendix 6 of BS 7671.
© Copyright The Electrical Safety Council (July 2006)

Please see the 'Notes for Recipients' on the reverse of this page.

IPM3/5

**Supply characteristics, earthing and bonding arrangements**

**Supply characteristics** Nominal voltage of the supply.

**System type** TT, TNS or TNCS.

**Nominal frequency** Normally 50 Hz.

**Prospective fault current** Is the highest current that could flow within the installation between live conductors, or live conductors and earth. This should be measured or obtained by enquiry. If it is measured, remember that on a 3 phase system the value between phase and neutral must be doubled.

**External earth loop impedance, $Z_e$** This is the external earth fault loop impedance measured between the phase and earthing conductor for the installation.

### Characteristics of the supply protective device

**BS type** Can normally be found printed on the service head.

**Nominal current rating** Can normally be found printed on the service head.

**Short circuit capacity** This will depend on the type, but if in doubt reference should be made to Table 7.4 in the *On-Site Guide*.

**Main switch or circuit breaker** The Type is normally printed on it but reference can be made to Appendix 2 of BS 7671 if required.

**Number of poles** Does the switch break all live conductors when opened, or is it single pole only?

**Supply conductor material and size** This refers to the meter tails.

**Voltage rating** This will usually be printed on the device.

**Current rating** This will usually be printed on the device.

**RCD operating current, $I\Delta n$** This is the trip rating of the RCD and should only be recorded if the RCD is used as a main switch.

**RCD operating time at $I\Delta n$** Only to be recorded if the RCD is used as a main switch.

## Means of earthing

**Distributors** Facility or earth electrode?

**Type** If earth electrode.

**Electrode resistance** Usually measured as $Z_e$.

**Location** Where is the earth electrode?

**Method of measurement** Has an earth fault loop tester or an earth electrode tester been used to carry out the test? To do this test correctly, the earthing conductor should be disconnected to avoid the introduction of parallel paths. This will of course require isolation of the installation; in some instances this may not be practical or possible for various reasons.

If isolation is not possible, the measurement should still be carried out to prove that the installation has an earth. The measured value of $Z_e$ should be equal to or less than any value for $Z_e$ documented on previous test certificates. If the measurement is higher than those recorded before, then further investigation will be required.

The higher measurement could be caused by corrosion, a loose connection or damage.

If the means of earthing is by an earth electrode, the soil conditions may have changed. This would be considered normal providing that the measured value is less than 200Ω and the system is protected by a residual current device.

## Main protective conductors

### Earthing conductor

**Conductor material** What is it made of? Unless special precautions are taken in accordance with BS 7671, this should be copper.

**Conductor cross-sectional area** This must comply with Regulation section 543. If the system is PME then Regulation 544.1. In most domestic installations this will require the size to be 16mm$^2$. Further information can be found in Table 4.1 of the *On-Site Guide*.

**Continuity check** This requires a tick only and is usually a visual check, provided that the conductor is visible in its entirety.

## Main equipotential bonding conductors

**Conductor material** What is it made of? Unless special precautions are taken in accordance with BS 7671 this should be copper.

**Conductor cross-sectional area** This must comply with Regulation 544. In most domestic installations the required size will be $10mm^2$. Further information is available in Table 10A of the *On-Site Guide*.

**Bonding of extraneous conductive parts** All services, structural steel, lightning conductors and central heating systems should be equipotential bonded. See Regulation 411.3.1.2 or Chapter 4 of the *On-Site Guide*. Normally a tick required.

### SUMMARY OF THE INSPECTION

This section of the certificate is to detail the overall condition of the installation.

It is often easier for the inspector to break the installation into specific areas, for instance:

- Any change of use or environment which may have had an affect on the installation
- Earthing arrangements
- Bonding
- Isolation
- Age
- Safety

**Overall assessment**. This will either be satisfactory or unsatisfactory. In general terms, if the observation area of the form has any defects other than code 4, the assessment must be unsatisfactory.

## Safety Certificate for Periodic Inspection

### NOTES FOR RECIPIENT

**THIS SAFETY CERTIFICATE IS AN IMPORTANT AND VALUABLE DOCUMENT
WHICH SHOULD BE RETAINED FOR FUTURE REFERENCE**

**This safety certificate has been issued to confirm that the electrical installation work to which it relates has been designed, constructed, inspected, tested and verified in accordance with the national standard for the safety of electrical installations, British Standard 7671 (as amended) - *Requirements for Electrical Installations* (formerly known as the IEE Wiring Regulations).**

**Where, as will often be the case, the installation incorporates a residual current device (RCD), there should be a notice at or near the main switchboard or consumer unit stating that the device should be tested at quarterly intervals. For safety reasons, it is important that you carry out the test regularly.**

**Also for safety reasons, the complete electrical installation will need to be inspected and tested at appropriate intervals by a competent person. The maximum interval recommended before the next inspection is stated on Page 2 under *Next Inspection*. There should be a notice at or near the main switchboard or consumer unit indicating when the inspection of the installation is next due.**

This report is intended for use by electrical contractors not enrolled with NICEIC or by NICEIC Approved Contractors working outside the scope of their enrolment. The certificate consists of at least five numbered pages.

For installations having more than one distribution board or more circuits than can be recorded on pages 4 and 5, one or more additional *Schedules of Circuit Details for the Installation*, and *Schedules of Test Results for the Installation* (pages 6 and 7 onwards) should form part of the certificate.

This certificate is intended to be issued only for a new electrical installation or for new work associated with an alteration or addition to an existing installation. It should not have been issued for the inspection of an existing electrical installation. A 'Periodic Inspection Report' should be issued for such a periodic inspection.

You should have received the certificate marked 'Original' and the electrical contractor should have retained the certificate marked 'Duplicate'.

**If you were the person ordering the work, but not the user of the installation, you should pass this certificate, or a full copy of it including these notes, the schedules and additional pages (if any), immediately to the user.**

**The 'Original' certificate should be retained in a safe place and shown to any person inspecting or undertaking further work on the electrical installation in the future. If you later vacate the property, this certificate will demonstrate to the new user that the electrical installation complied with the requirements of the national electrical safety standard at the time the certificate was issued.**

Page 1 of this certificate provides details of the electrical installation, together with the name(s) and signature(s) of the person(s) certifying the three elements of installation work: design, construction and inspection and testing. Page 2 identifies the organisation(s) responsible for the work certified by their representative(s).

Certification for inspection and testing provides an assurance that the electrical installation work has been fully inspected and tested, and that the electrical work has been carried out in accordance with the requirements of BS 7671 (except for any departures sanctioned by the designer and recorded in the appropriate box(es) of the certificate).

If wiring alterations or additions are made to an installation such that wiring colours to two versions of BS 7671 exist, a warning notice should have been affixed at or near the appropriate consumer unit.

## Safety Certificate for Electrical Installation

## NOTES FOR RECIPIENTS

### THIS SAFETY CERTIFICATE IS AN IMPORTANT AND VALUABLE DOCUMENT
### WHICH SHOULD BE RETAINED FOR FUTURE REFERENCE

The purpose of periodic inspection is to determine, so far as is reasonably practicable, whether an electrical installation is in a satisfactory condition for continued service. This report provides an assessment of the condition of the electrical installation identified overleaf at the time it was inspected, taking into account the stated extent of the installation and the limitations of the inspection and testing.

The report has been issued in accordance with the national standard for the safety of electrical installations, British Standard 7671 (as amended) - *Requirements for Electrical Installations* (formerly known as the IEE Wiring Regulations).

Where the installation incorporates a residual current device (RCD), there should be a notice at or near the main switchboard or consumer unit stating that the device should be tested at quarterly intervals. For safety reasons, it is important that you carry out the test regularly.

Also for safety reasons, the electrical installation will need to be re-inspected at appropriate intervals by a competent person. The recommended maximum time interval to the next inspection is stated on page 3 in Section I (*Next Inspection*). There should be a notice at or near the main switchboard or consumer unit indicating when the next inspection of the installation is due.

The report consists of at least six numbered pages. The report is invalid if any of the pages identified in Section H are missing.

For installations having more than one distribution board or more circuits than can be recorded on Pages 5 and 6, one or more additional *Schedules of Circuit Details for the Installation*, and *Schedules of Test Results for the Installation* (pages 7 and 8 onwards) should form part of the report.

This report is intended to be issued only for the purpose of reporting on the condition of an existing electrical installation. The report should identify, so far as is reasonably practicable and having regard to the extent and limitations recorded in Section D, any damage, deterioration, defects, dangerous conditions and any non-compliances with the requirements of the national standard for the safety of electrical installations which may give rise to danger. It should be noted that the greater the limitations applying to a report, the less its value.

The report should not have been issued to certify that a new electrical installation complies with the requirements of the national safety standard. An 'Electrical Installation Certificate' or a 'Domestic Electrical Installation Certificate' (where appropriate) should be issued for the certification of a new installation.

You should have received the report marked 'Original' and the electrical contractor should have retained the report marked 'Duplicate'.

If you were the person ordering the work, but not the user of the installation, you should pass this report, or a full copy of it including these notes, the schedules and additional pages (if any), immediately to the user.

The 'Original' report form should be retained in a safe place and shown to any person inspecting or undertaking further work on the electrical installation in the future. If you later vacate the property, this report will provide the new user with an assessment of the condition of the electrical installation at the time the periodic inspection was carried out.

Section D addresses the extent and limitations of the report by providing boxes for the *Extent of the electrical installation covered by this report* and the *Agreed limitations, if any, on the inspection and testing*. Information

## NOTES FOR RECIPIENT
### (continued from the reverse of page 1)

Where responsibility for the *design*, the *construction* and the *inspection and testing* of the electrical work is divided between the electrical contractor and one or more other bodies, the division of responsibility should have been established and agreed before commencement of the work. In such a case, the absence of certification for the *construction*, or the *inspection and testing* elements of the work would render the certificate invalid. If the *design* section of the certificate has not been completed, you should question why those responsible for the design have not certified that this important element of the work is in accordance with the national electrical safety standard.

All unshaded boxes should have been completed either by insertion of the relevant details or by entering 'N/A', meaning 'Not Applicable', where appropriate.

Where the electrical work to which this certificate relates includes the installation of a fire alarm system and/or an emergency lighting system (or a part of such systems) in accordance with British Standards BS 5839 and BS 5266 respectively, this electrical safety certificate should be accompanied by a separate certificate or certificates as prescribed by those standards.

Where the installation can be supplied by more than one source, such as the public supply and a standby generator, the number of sources should have been recorded in the box entitled Number of Supplies, under the general heading *Supply Characteristics and Earthing Arrangements* on page 2 of the certificate, and the *Schedule of Test Results* compiled accordingly. Where a number of sources are available to supply the installation, and where the data given for the primary source may differ from other sources, an additional page should have been provided which gives the relevant information relating to each additional source, and to the associated earthing arrangements and main switchgear.

# GUIDANCE FOR RECIPIENTS ON THE RECOMMENDATION CODES

**Only one Recommendation Code should have been given for each recorded observation.**

**Recommendation Code 1**

**Where an observation has been given a Recommendation Code 1 (requires urgent attention), the safety of those using the installation may be at risk.**

The person responsible for the maintenance of the installation is advised to take action without delay to remedy the observed deficiency in the installation, or to take other appropriate action (such as switching off and isolating the affected part(s) of the installation) to remove the potential danger. The electrical contractor issuing this report will be able to provide further advice.

**It is important to note that the recommendation given at Section I *Next Inspection* of this report for the maximum interval until the next inspection, is conditional upon all items which have been given a Recommendation Code 1 being remedied without delay.**

**Recommendation Code 2**

Recommendation Code 2 (requires improvement) indicates that, whilst the safety of those using the installation may not be at immediate risk, remedial action should be taken as soon as possible to improve the safety of the installation to the level provided by the national standard for the safety of electrical installations, BS 7671. The electrical contractor issuing this report will be able to provide further advice.

Items which have been attributed Recommendation Code 2 should be remedied as soon as possible (see Section F).

**Recommendation Code 3**

Where an observation has been given a Recommendation Code 3 (requires further investigation), the inspection has revealed an apparent deficiency which could not, due to the extent or limitations of this inspection, be fully identified. Items which have been attributed Recommendation Code 3 should be investigated as soon as possible (see Section F).

The person responsible for the maintenance of the installation is advised to arrange for the further examination of the installation to determine the nature and extent of the apparent deficiency.

**Recommendation Code 4**

Recommendation Code 4 [does not comply with BS 7671 (as amended)] will have been given to observed non-compliance(s) with the **current** safety standard which do not warrant one of the other Recommendation Codes. It is not intended to imply that the electrical installation inspected is unsafe, but careful consideration should be given to the benefits of improving these aspects of the installation. The electrical contractor issuing this report will be able to provide further advice.

# Safety in electrical testing

## CORRECT SELECTION OF PROTECTIVE DEVICES

(*While protective devices are mentioned throughout this book, this chapter brings all of the information together for reference*)

When carrying out an inspection and test on any electrical installation it is important to ensure that the correct size and type of device has been installed.

To do this we must have a good knowledge of the selection of protective devices and the type of circuits that they are protecting.

### Why are they installed?

Protective devices are installed to protect the cable of the circuit from damage – this could be caused by overload, <u>overcurrent and fault current</u>.

The definition for overload given in Part 2 of BS 7671:2004 is: *Overcurrent occurring in a circuit which is electrically sound.* This is when the circuit is installed correctly and the equipment connected to it is drawing too much current. For instance:

An electric motor connected to the circuit is used on too heavy a load, leading to an overload of the circuit. Provided that the correct size of protective device was installed, the device will operate and interrupt the supply preventing the cable from overloading.

If additional luminaries were installed on an existing circuit which was already fully loaded, the protective device should operate and protect the cable of the circuit.

Overcurrent is a current flow in a circuit which is greater than the rated current carrying capacity of the cables. This would normally be due to a fault on the circuit or incorrect cable selection. For example:

> If a 20 amp cable protected by a 32 amp circuit breaker was loaded by 25 A, the cable would overheat and the device would continue to allow current to flow – this could damage the cable.

A fault current is a current which is flowing in a circuit due to a fault. For example:

> A nail is driven through a cable causing an earth fault or a short circuit fault. This would cause a very high current to flow through the circuit, which must be interrupted before the conductors reach a temperature that could damage the insulation or even the conductors.

So what are we looking for with regard to protective devices during an inspection?

### What type of device is it? Is it a fuse or circuit breaker?

A fuse has an element which melts when too much current is passed through it, whether by overload or fault current.

Fuses in common use are:

> BS 3036 semi-rewirable fuse
> BS 88 cartridge fuse
> BS 1361 cartridge fuse

A circuit breaker is really two devices in one unit. The overload part of the device is a thermal bi metal strip, which heats up when a current of a higher value than the nominal current rating ($I_n$) of the device passes through it.

Also incorporated within the device is a magnetic trip, which operates and causes the device to trip when a fault current flows through it. For the device to operate correctly it must operate within 0.1 seconds. The current which has to flow to operate the device in the required time has the symbol ($I_a$).

Circuit breakers in common use are:

☞ BS 3871 types 1, 2 and 3
☞ BS EN types B, C and D (*A is not used, this is to avoid confusion with Amps*)

*Is the device being used for protection against indirect contact?*
*In most instances this will be the case.*

*What type of circuit is the device protecting, is it supplying fixed equipment only, or could it supply handheld equipment?*

If the circuit supplies fixed equipment only, the device must operate on fault current within 5 seconds. If it supplies socket outlets it must operate within 0.4 seconds and usually be RCD protected (*see BS 7671, Regulations 411.3.2 and 413.32.3*).

When using circuit breakers to BS 3871 and BS EN 60898 these times can be disregarded. Providing the correct $Z_S$ values are met, they will operate in 0.1 seconds or less.

*If it is a circuit breaker is it the correct type?*

Table 7.2b of the *On-Site Guide* provides a good reference for this.

Types 1 and B should be used on circuits having only resistive loads (*have you ever plugged in your 110v site transformer and found that it operated the circuit breaker? If you have it will be because it was a type 1 or B*).

Types 2, C and 3 should be used for inductive loads such as fluorescent lighting, small electric motors and other circuits, where surges could occur.

Types 4 and D should be used on circuits supplying large transformers or any circuits where high inrush currents could occur.

*Will the device be able to safely interrupt the prospective fault current which could flow in the event of a fault?*

Table 7.2A or the manufacturer's literature will provide information on the rated short circuit capacity of protective devices.

*Is the device correctly coordinated with the load and the cable?*

Correct coordination is:

> Current carrying capacity of the cable under its installed conditions must be equal to or greater than the rated current of the protective device ($I_z$).

The rated current carrying capacity of the protective device ($I_n$) must be equal to or greater than the design current of the load ($I_b$).

In short, $I_z > I_n > I_b$ (*Appendix 4, item 4 BS 7671 or Appendix 6 On-Site Guide*).

### Additional information regarding circuit breakers

*Overload current*

The symbol for the current required to cause a protective device to operate within the required time on overload is ($I_2$).

Circuit breakers with nominal ratings up to 60 amps must operate within 1 hour at 1.45 × their nominal rating ($I_n$).

Circuit breakers with nominal ratings above 60 amps must operate within 2 hours at 1.45 × its rating ($I_n$).

At 2.55 times the nominal rating ($I_n$), circuit breakers up to 32 amps must operate within 1 minute; and circuit breakers above 32 amps, must operate within 2 minutes.

They must <u>not</u> trip within 1 hour at up to 1.13 their nominal rating ($I_n$).

*Maximum earth fault loop impedance values ($Z_S$) for circuit breakers*

These values can be found in Table 41.3 in BS 7671.

Because these devices are required to operate within 0.1 of a second, they will satisfy the requirements of BS 7671 with regard to disconnection times in all areas. Therefore, the $Z_S$ values for these devices are the same wherever they are to be used (this only applies to circuit breakers) even in special locations where the disconnection time must be 0.2 seconds.

*Calculation of the Maximum $Z_S$ of circuit breakers*

It is often useful to be able to calculate the maximum $Z_S$ value for circuit breakers without the use of tables. This is quite a simple process for BS 3871 and BS EN 60898 devices. Let's use a 20A BS EN 60898 device as an example:

Table 7.2B shows that a type B device must operate within a window of 3 to 5 times its rating. As electricians we always look at the worst case scenario. Therefore, we must assume that the device will not operate until a current equal to 5 times its rating flows through it ($I_a$). For a 20 A type B device this will be $5 \times 20\,A = 100\,A$.

tested, then the test voltage must be 250 V d.c. and the maximum resistance value is 0.5 MΩ – although this would be considered a very low value and any value below 5 MΩ must be investigated.

For a test between the actual transformer windings the test voltage is increased to 500 V d.c. The minimum insulation resistance value is 1 MΩ although any value below 5 MΩ should be investigated.

## TESTING A 3 PHASE INDUCTION MOTOR

There are many types of 3 phase motors but by far the most common is the induction motor. It is quite useful to be able to test them for serviceability.

Before carrying out electrical tests it is a good idea to ensure that the rotor turns freely. This may involve disconnecting any mechanical loads. The rotor should rotate easily and you should not be able to hear any rumbling from the motor bearings. Next, if the motor has a fan on the outside of it, check that it is clear of any debris which may have been sucked in to it. Also check that any air vents into the motor are not blocked.

Generally, if the motor windings are burnt out there will be an unmistakable smell of burnt varnish. However, it is still a good idea to test the windings as the smell could be from the motor being overloaded. Three phase motors are made up of three separate windings – in the terminal box there will be six terminals as each motor winding will have two ends. The ends of the motor windings will usually be identified as W1, W2; U1, U2; or V1, V2. The first part of the test is carried out using a low resistance ohm meter. Test each winding end to end (W1 to W2, U1 to U2 and V1 to V2). The resistance of each winding should be approximately the same and the resistance value will depend on the size of the motor. If the resistance values are different, then the motor will not be electrically balanced and it should be sent for rewinding. If resistance values are the same, then the next test is carried out using an insulation resistance tester. Join W1 and W2 together, U1 and U2 together and V1 and V2 together. Carry out an insulation resistance test between the joined ends, i.e. W to U then W to V and then between U and V. Then repeat the test between joined ends and the case, or the earthing terminal of the motor (*these tests can be in any order to suit you*). Providing the insulation resistance is 2 MΩ or greater then the motor is fine. If the insulation resistance is above 0.5 MΩ this could be due to dampness and it is often a good idea to run the motor for a while before carrying out the insulation test again as the motor may dry out with use.

To reconnect the motor windings in star, join W2, U2 and V2 together and connect the 3 phase motor supply to W1, U1 and V1. If the motor rotates in the wrong direction, swap two of the phases of the motor supply.

To reconnect the motor windings in delta, join W1 to U2, U1 to V2 and V1 to W2 and then connect the 3 phase motor supply one to each of the joined ends. If the motor rotates in the wrong direction, swap two phases of the motor supply.

# Appendix
# A

# IP codes

## INGRESS PROTECTION

In BS 7671 Wiring Regulations the definition of an enclosure is 'A Part providing protection of equipment against certain external influences and in any direction against direct contact'.

To ensure that we use the correct protection to suit the environment where the enclosure is installed, codes are used. These codes are called IP codes. IP stands for ingress protection and is an international classification system for the sealing of electrical enclosures or equipment.

The system uses the letters IP followed by two or three digits. The first digit indicates the degree of protection required for the intrusion of foreign bodies such as dust, tools and fingers.

The second digit provides an indication of the degree of protection required against the ingress of moisture.

If a third digit is used, a letter would indicate the level of protection against access to hazardous parts by persons; a number would indicate the level of protection against impact.

Where an 'X' is used, it is to show that nothing is specified. For example, if a piece of equipment is rated at IPX8, it would require protection to allow it to be submersed in water. Clearly if a piece of equipment can be submersed safely, then dust will not be able to get in to it and no protection against the ingress of dust would be required.

## Table of Ip ratings

| Dust and foreign bodies | Level of Protection | Moisture | Level of Protection |
|---|---|---|---|
| 0 | No special protection | 0 | No special protection |
| 1 | 50 mm | 1 | Dripping water |
| 2 | 12.5 mm diameter and 80 mm long (finger) | 2 | Dripping water when tilted at 15° |
| 3 | 2.5 mm | 3 | Rain proof |
| 4 | 1 mm | 4 | Splash proof |
| 5 | Limited dust | 5 | Sprayed from any angle (jet proof) |
| 6 | Dust tight | 6 | Heavy seas and powerful jets |
| | | 7 | Immersion up to 1 M |
| | | 8 | Submersion 1 M + |

## Third digit, usually a letter

| | |
|---|---|
| A | The back of a hand or 50 mm sphere |
| B | Standard finger 80 mm long |
| C | Tool 2.5 mm diameter, 100 mm long, must not contact hazardous areas |
| B | Wire 1 mm diameter, 100 mm long, must not contact hazardous areas |

The third number for impact is not used in BS 7671 and is not included in this book.

Supply system is TN-S 230 volt measured $Z_S$ is 0.43 $\Omega$, and PFC is 1.2 kA.

All circuits are protected by BS EN 60898 type B circuit breaker.

### EXERCISE 5

Use BS 7671 to find the regulation number indicating where RCDs should be used and what trip rating they should have for:

- Fire protection on farms
- Protection where flammable materials are stored
- Fixed equipment in Zones 1–3 in bathrooms
- Restrictive conductive locations
- TT systems
- Circuits with a high earth fault loop impedance ($Z_S$)
- Caravan parks
- Sockets likely to supply portable equipment used out doors
- Swimming pools

# EXERCISE 6

| Circuit Description | Overcurrent Device | | Wiring Conductors | | Length | $R_1 + R_2$ | Rn | Max $Z_s$ | Actual $Z_s$ | Voltage Drop | Insulation resistance | |
|---|---|---|---|---|---|---|---|---|---|---|---|---|
| | Type | Rating | Live | CPC | | | | | | | Live/Live | Live/CPC |
| 1. Shower | B | 50 A | 10.0 mm | 4.0 mm | 18 m | | | | | | >200 MΩ | >200 MΩ |
| 2. Cooker | B | 32 A | 6.00 mm | 2.5 mm | 21 m | | | | | | 56 MΩ | >200 MΩ |
| 3. Ring Circuit | B | 32 A | 2.5 mm | 1.5 mm | 66 m | | | | | | 110 MΩ | >200 MΩ |
| 4. Ring Circuit | B | 32 A | 2.5 mm | 1.5 mm | 83 m | | | | | | 80 M | >200 MΩ |
| 5. Radial sockets | B | 20 A | 2.5 mm | 1.5 mm | 38 m | | | | | | 60 MΩ | >200 MΩ |
| 6. Immersion Heater | B | 16 A | 2.5 mm | 1.5 mm | 12 m | | | | | | >200 MΩ | >200 MΩ |
| 7. Lighting | C | 6 A | 1.5 mm | 1.00 mm | 43 m | | | | | | >200 MΩ | >200 MΩ |
| 8. Lighting | B | 6 A | 1.00 mm | 1.00 mm | 48 m | | | | | | 100 MΩ | >200 MΩ |
| 9. Lighting | B | 6 A | 1.00 mm | 1.00 mm | 58 m | | | | | | 120 MΩ | >200 MΩ |
| 10. Supply to Shed | C | 16 A | 4.00 mm | 1.5 mm | 23 m | | | | | | >200 MΩ | >200 MΩ |

Using the values given in BS 7671 complete the chart. Calculate PFC. Calculate the maximum volt drop for each circuit. Calculate the insulation resistance for the complete installation between live conductors and live conductors and earth. Indicate any non-compliances.

# Appendix C

# Questions

## Test 1

1. An insulation resistance test has been carried out on a 6-way consumer's unit. The circuits recorded values of $5.6\,\text{m}\Omega$, $8.7\,\text{m}\Omega$, $>200\,\text{m}\Omega$, $>200\,\text{m}\Omega$, $12\,\text{m}\Omega$ and $7\,\text{m}\Omega$.
   Calculate the total resistance of the installation and state giving reasons whether or not the installation is acceptable.

2. A ring circuit is 54 metres long and is wired in $2.5\,\text{mm}^2/1.5\,\text{mm}^2$ thermoplastic cable. The protective device is a 32 A BS EN 60898 type C device and the $Z_e$ for the installation is $0.24\Omega$. The resistance of $2.5\,\text{mm}^2$ copper is $7.41\,\text{m}\Omega$ per metre and $1.5\,\text{mm}^2$ copper is $12.1\,\text{m}\Omega$ per metre.
   (i)  Calculate the $Z_S$ for the circuit.
   (ii) Will the protective device be suitable?

3. An A2 radial circuit is wired in $4\,\text{mm}^2$ thermoplastic twin and earth cable. It is 23 metres long. The circuit has on it four twin 13 amp socket outlets. Protection is by a 30 A BS 3036 semi-enclosed fuse. $Z_e$ for the installation is $0.6\Omega$.
   Socket 1 is 12 metres from the consumer unit, socket 2 is 6 metres from socket 1, and socket 3 is 2.5 metres from socket 2.
   (i)  Calculate the $R_1 + R_2$ value at each socket outlet.
   (ii) Will the circuit protective device be suitable?

4. A 9.5 kW electric shower has been installed using $10\,\text{mm}^2/4\,\text{mm}^2$ thermoplastic twin and earth cable which is 14.75 metres long. The circuit is to be connected to a spare way in the existing consumer's unit. Protection is by a BS 3036 semi-enclosed 45 amps rewirable fuse. $Z_e$ for the system is $0.7\Omega$. The temperature at the time of testing is 20°C.
   (i)  Calculate $R_1 + R_2$ for this circuit.
   (ii) Will this circuit meet the required disconnection time?

5. A ring circuit is wired $4 \, mm^2$ singles in conduit. The circuit is 87 metres in length and is protected by a 32 A BS 3871 type 2 circuit breaker, the maximum $Z_S$ permissible is $1.07 \, \Omega$ and the actual $Z_e$ is $0.63 \, \Omega$. Calculate:
   (i) The expected $Z_S$.
   (ii) The maximum permissible length that could be allowed for a spur in $4 \, mm^2$ cable.

6. List the certification that would be required after the installation of a new lighting circuit.

7. List three non-statutory documents relating to electrical installation testing.

8. List four reasons why a Periodic Test Report would be required.

9. Apart from a new installation, under which circumstances would a Periodic Inspection Report **not** be required?

10. A ring circuit is wired in $2.5 \, mm^2 / 1.5 \, mm^2$. The resistance of the phase and neutral loops were each measured at $0.3 \, \Omega$. Calculate the:
    (i) The resistance between P and N at each socket after all interconnections have been made.
    (ii) End to end resistance of the CPC.
    (iii) Resistance between P and CPC at each socket after all interconnections have been made.

11. A spur has been added to the ring circuit in Question 10. The additional length of cable used is 5.8 metres.
    Calculate the $R_1 + R_2$ for this circuit.

12. What is a 'statutory' document?

13. What is a 'non-statutory document'?

14. Why is it important to carry out testing on a new installation in the correct sequence?

15. How many special locations are listed in the BS 7671 amended to 2004?

16. State the affect that increasing the length of a conductor could have on its insulation resistance.

17. An installation has seven circuits. Circuits 1, 4 and 6 have insulation resistances of greater than $200 \, m\Omega$. Circuits 2, 3, 5 and 7 have resistance values of 50, 80, 60, and 50, respectively. Calculate the total resistance of the circuit.

18. State the correct sequence of tests for a new domestic installation connected to a TT supply.

19. List, in the correct sequence, the instruments required to carry out the tests in Question 18.

20. State the values of the test currents required when testing a 30 mA RCD used for supplementary protection against direct contact.

21. How many times its rated operating current is required to operate a type B BS EN 60898 circuit breaker instantaneously?
22. What is the maximum resistance permitted for equipotential bonding?
23. What would be the resistance of 22 metres of a single $10\,mm^2$ copper conductor?
24. Which type of supply system uses the mass of earth for its earth fault return path?
25. The Table below shows the resistance values recorded at each socket on a ring circuit during a ring circuit test after the interconnections had been made. Are the values as expected? If not, what could the problem be? The temperature is 20°C; the end to end resistances of the conductors are: Phase $0.45\,\Omega$, neutral $0.46\,\Omega$ and CPC is $0.75\,\Omega$.

| Socket | P to N | P to CPC |
| --- | --- | --- |
| 1 | 0.225 | 0.35 |
| 2 | No reading | No reading |
| 3 | 0.224 | No reading |
| 4 | No reading | 0.35 |
| 5 | 0.34 | 0.50 |
| 6 | 0.4 | 0.35 |
| 7 | 0.22 | 0.35 |

26. A lighting circuit is to be wired in $1\,mm^2$ twin and earth thermoplastic cable, the circuit is protected by a 5 A BS 3036 fuse. What would be the maximum length of cable permissible to comply with the earth fault loop impedance requirements ($Z_S$)? ($Z_e$ is 0.45.)
27. With regard to the *On-Site Guide* what are the stated earth loop impedance values outside of a consumer's installation for a TT, TNS, and TNCS supply?
28. A ring final circuit has twelve twin 13 amp socket outlets on it. How many unfused spurs would it be permissible to add to this circuit?
29. How many fused spurs would it be permissible to connect to the ring circuit in Question 28?
30. Name the document that details the requirements for electrical test equipment.

31. State three extraneous conductive parts that could be found within a domestic installation.
32. State four exposed conductive parts commonly found within an electrical installation.
33. State the minimum c.s.a. for a non-mechanically protected, supplementary bonding conductor that could be used in a bathroom.
34. What is the minimum acceptable insulation resistance value permissible for a complete 400 V a.c. 50 Hz installation?
35. State the test voltage and current required for an insulation test carried out on a 230 V a.c. 50 Hz installation.
36. The Electricity at Work Regulations state that for a person to be competent when carrying out inspecting and testing on an electrical installation they must be ............ What?
    (i)   In possession of technical knowledge.
    (ii)  Experienced, or
    (iii) Supervised.
37. A 400 V a.c. installation must be tested with an insulation resistance tester set at ......... volts.
38. List three requirements of GS 38 for test leads.
39. List three requirements of GS 38 for test probes.
40. Name a suitable piece of equipment that could be used for testing for the presence of voltage while carrying out the isolation procedure.
41. To comply with BS 7671 the purpose of the initial verification is to verify that...............
42. To comply with GN 3, what are the four responsibilities of the inspector?
43. When testing a new installation, a fault is detected on a circuit. State the procedure that should be carried out.
44. State three reasons for carrying out a polarity test on a single phase installation.
45. On which type of ES lampholders is it **not** necessary to carry out a polarity test?
46. What is the minimum requirement of BS 7671 for ingress protection of electrical enclosures?
47. A 6 A Type B circuit breaker trips each time an earth loop impedance test is carried out on its circuit. How could the $Z_S$ value for this circuit be obtained?
48. What is the maximum rating permissible for before a motor would require overload protection?

49. Identify the type of circuit breaker that should be used for:
    (a) Discharge lighting in a factory.
    (b) A large transformer.
    (c) A3 phase motor.
50. Identify three warning labels and notices that could be found in an installation.
51. The circuits in the table have been tested and the earth fault loop impedance values for each circuit are as shown. Using the rule-of-thumb method to identify whether the circuits will comply with BS 7671.

| Measured $Z_S$ | Maximum $Z_S$ |
| --- | --- |
| 0.86 Ω | 1.2 Ω |
| 0.68 Ω | 0.96 Ω |
| 1.18 Ω | 1.5 Ω |
| 2.8 Ω | 4 Ω |
| 1.75 Ω | 2.4 Ω |

52. State the minimum IP rating for fixed equipment in Zone 2 of a bathroom.
53. State the minimum size for an equipotential bonding conductor installed in a TNS system with $25\,mm^2$ metre tails.
54. Identify the documentation that should be completed after the installation of a cooker circuit.
55. State four non-statutory documents.
56. State four statutory documents.
57. State the sequence of colours for a new 3 phase and neutral system.


## Test 2

1. List the items of documentation required when a consumer unit is changed.
2. Why are records of inspection and test necessary?
3. Name the statutory document which requires records of an installation to be kept.
4. How would you define competence?
5. What is the reason for carrying out periodic testing?

6. State three ways of achieving safety when using test equipment.
7. State the tests which would record readings in MΩ, kA and mS.
8. List the live tests required on a TNS system.
9. Main equipotential bonding must be connected to the MET and which extraneous conductive parts?
10. What would affect the resistance of a copper conductor.
11. Which method should be used to measure the continuity of a main equipotential bonding conductor?
12. Why should care be taken during a test of equipotential bonding?
13. Above which value is equipotential bonding unacceptable?
14. Why is a ring circuit test carried out?
15. The CPC and Phase conductors of a ring are interconnected for testing, when the test is carried out the reading increases significantly towards the centre of the ring. If it only increased slightly what would be the reason?
16. Between which conductors should insulation resistance tests be carried out?
17. Before an insulation test is carried out what should be checked?
18. What method is used to compensate for conductor operating and ambient temperature when an earth loop impedance test is carried out?
19. Before carrying out an RCD test it is important to carry out an earth fault loop impedance test. Why?
20. Define prospective short circuit current.
21. Define prospective earth fault current.
22. Which value of fault current should be recorded on a test certificate, PSCC or PEFC?
23. When would it be acceptable to not carry out supplementary bonding within a bathroom?
24. List three recognized types of earth electrode.
25. Final circuits in a consumer unit must contain protection against?


## Test 3

1. What is the status of an electrician who will be carrying an inspection of an installation?
2. State **two** areas of responsibility for an inspector as listed in GN3?
3. State the main difference between a non statutory document and a statutory document.
4. List an example of a:
   (i) Non-statutory document.
   (ii) Statutory document.

# Appendix
# D

# Answers to exercises

(*The Certificates can be found at the end of the chapter*)

## EXERCISE 1

1. and **2.**   Part P – Domestic Electrical Installation Certificate – Part 2.
3. Part P – Domestic Electrical Installation Certificate.
4. $1.92 \times 0.75 = 1.44\ \Omega$. This is the maximum permissible $Z_S$ for a BS 88 32amp fuse. As the measured value of $Z_S$ was $0.48\ \Omega$ this circuit is fine. The measured $Z_S$ is lower due to parallel paths through the bonding conductors, etc.
5. Minor Domestic Electrical Installation Works Certificate.
6. Use an earth fault loop impedance test instrument set on loop, plug lead into socket outlet on cooker control unit and take earth loop impedance reading.

## EXERCISE 2

1. Part P – Domestic Electrical Installation Certificate; Part P – Domestic Electrical Installation Certificate – Part 2.
2. Minor Domestic Electrical Installation Works Certificate.

EXERCISE 3

1. Rule-of-thumb

| Circ no. | Max $Z_S$ | Calculated | Actual | $\times / \checkmark$ |
|---|---|---|---|---|
| 1 | $1.09 \times 0.8 =$ | 0.87 | 0.82 | ok |
| 2 | $1.09 \times 0.8 =$ | 0.87 | 0.81 | ok |
| 3 | $1.09 \times 0.8 =$ | 0.87 | 0.83 | ok |
| 4 | $2.55 \times 0.8 =$ | 2.73 | 0.94 | ok |
| 5 | $9.58 \times 0.8 =$ | 7.7 | 1.95 | ok |
| 6 | $9.58 \times 0.8 =$ | 7.7 | 2.64 | ok |

2. Periodic inspection and test reports. Periodic Inspection Report for an Electrical Installation; Supply Characteristics and Earthing Arrangements; Survey and Test Report Schedule.
   (a) There is no supplementary bonding in the bathroom.
   (b) The earthing conductor is undersized.
   (c) RCD protection is advisable for the ground floor ring circuit although it is not a requirement as the wiring regulations are not retrospective.

The inspection checklist indicates '$\times$' for presence of RCD for supplementary protection as there is a downstairs ring circuit which should have one to comply with current regulations, this also requires a '$\times$' for particular protective measures for special locations.

Electrical separation has a tick as there is a shaver socket in the bathroom.

A tick $\checkmark$ is required in segregation of band I and II circuits where the installation has circuits such as bell, telephone and TV aerials, etc.

EXERCISE 4

Part P – Domestic Electrical Installation Certificate – Part 2.

# DOMESTIC ELECTRICAL INSTALLATION CERTIFICATE

Issued in accordance with *British Standard 7671 – Requirements for Electrical Installations* by a Domestic Installer registered with NICEIC.

Warwick House, Houghton Hall Park, Houghton Regis, Dunstable, LU5 5ZX

**DOMESTIC INSTALLER**

**Original** (To the person ordering the work)

This certificate is not valid if the serial number has been defaced or altered

**DCP4/**

This safety certificate is an important and valuable document which should be retained for future reference

## DETAILS OF THE CLIENT

Client and address:

MRS F. G. GRANT
2 BISHOPS CLOSE
BATH
SOMERSET

Postcode SO3 6HT

## DETAILS OF THE INSTALLATION

Extent of the installation work covered by this certificate:

COOKER CIRCUIT

## DESIGN, CONSTRUCTION, INSPECTION AND TESTING

I/We being the person(s) responsible for the design, construction, inspection and testing of the electrical installation (as indicated by my/our signatures adjacent), particulars of which are described above, having exercised reasonable skill and care when carrying out the design, construction, inspection and testing, hereby CERTIFY that the said work for which I/we have been responsible is to the best of my/our knowledge and belief, in accordance with BS 7671, 2008 amended to ................ (date) except for the departures, if any, detailed as follows:

Details of departures from BS 7671, as amended (Regulations 120.3, 120.4)

NONE

## PARTICULARS OF THE DOMESTIC INSTALLER

Trading title: R AND T ELECTRICAL

Address:
6 COWBOY CLOSE
BATH
SOMERSET

Telephone No 01271 246810   Postcode SO3 6QT

NICEIC Registration No (Essential Information)  1 0 9 8 7 6 5 4 3

## ADDRESS OF THE INSTALLATION

Installation address:

2 BISHOPS CLOSE
BATH
SOMERSET

Postcode SO3 6HT

The installation is:

New ____
An addition ____
An alteration ____

The extent of liability of the signatory is limited to the work described above as the subject of this certificate

For the **DESIGN**, the **CONSTRUCTION** and the **INSPECTION AND TESTING** of the installation

Signature R Rogers   Name (CAPITALS) R. ROGERS   Date 1·07·08

The results of the inspection and testing reviewed by the **Qualified Supervisor**

Signature T Rigger   Name (CAPITALS) T. RIGGER   Date 2·07·08

## NEXT INSPECTION

I RECOMMEND that this installation is further inspected and tested after an interval of not more than §10YRS

§ Enter interval in terms of years, months or weeks, as appropriate

## COMMENTS ON EXISTING INSTALLATION

NONE

*Note: Enter NONE or, where appropriate, the page number(s) of additional page(s) of comments on the existing installation*

## SCHEDULE OF ADDITIONAL RECORDS*

NONE

*In the case of an alteration or addition see Section 633 of BS 7671*

See attached schedule

Please see the 'Notes for Recipients' on the reverse of this page.

Page 1 of 2

DCP4/1

* Where the electrical work to which this certificate relates includes the installation of a fire alarm system and/or an emergency lighting system (or a part of such systems), this electrical safety certificate should be accompanied by the particular certificate(s) for the system(s).

This form is based on the model Electrical Installation Certificate shown in Appendix 6 of BS 7671 (as amended)

Published by NICEIC Group Limited © Copyright The Electrical Safety Council (Jan 2008).

# DOMESTIC ELECTRICAL INSTALLATION CERTIFICATE

**DCP4/**

This certificate is not valid
if the serial number has
been defaced or altered

**Original** (To the person ordering the work)

## SUPPLY CHARACTERISTICS
*Tick boxes and enter details, as appropriate*

### System type(s)

**Number and type of live conductors**

| | |
|---|---|
| TN-S | ✓ |
| TN-C-S | |
| TT | |

| 1-phase (2 wire) | ✓ |
| 1-phase (3 wire) | |
| 3-phase (3 wire) | |
| 3-phase (4 wire) | |
| Other | Please state |

## Nature of supply parameters
*Notes: (1) by enquiry (2) by enquiry or by measurement (3) where more than one supply, record the higher or highest values*

Nominal voltage(s) $U_0^{(1)}$ ____ V
$U_0^{(1)}$ 230 V

Nominal frequency, $f^{(1)}$ 50 Hz

External earth fault, loop impedance, $Z_e^{(1)}$ 0.24 Ω

Prospective fault current, $I_{pf}^{(2)}$ 0.8 kA

**Single-phase** Prospective fault current, $I_{pf}^{(2)}$ 0.8 kA

**3-phase** Prospective fault current, $I_{pf}^{(2)}$ N/A kA

### Characteristics of primary supply overcurrent protective device(s)

BS(EN) 1361
Type 1
Rated current, $I_n$ 100 A

### Main switch or circuit-breaker

BS(EN) 60947-3
No. of poles 2
Supply conductors material Cu
Supply conductors csa 25 mm²

Short-circuit capacity 16.5 kA
Voltage rating 230 V
Rated current, $I_n$ 100 A
RCD operating current, $I_{\Delta n}^*$ 30 mA
RCD operating time (at $I_{\Delta n}$)* 43 ms

*\* applicable only where an RCD is used as a main circuit-breaker*

## PARTICULARS OF INSTALLATION AT THE ORIGIN
*Tick boxes and enter details, as appropriate*

### Means of earthing

| Distributor's facility | ✓ |
| Installation earth electrode | N/A |

### Details of installation earth electrode (where applicable)

Type (eg rod(s), tape etc) N/A
Location N/A
Electrode resistance, $R_A$ N/A Ω
Method of measurement N/A

### Earthing conductor

Conductor material Cu
Conductor csa 16 mm²
Continuity check ✓

## Main protective bonding conductors and bonding of extraneous-conductive-parts

Conductor material Cu
Conductor csa 10 mm²
Location (where not obvious)

Protective measures for fault protection ADS

| | |
|---|---|
| Water service | ✓ |
| Gas service | N/A |
| Oil service | N/A |
| Structural steel | N/A |
| Other incoming service(s) | N/A |

Maximum demand (load) 85 kW / Amps
*Delete as appropriate*
Number of smoke alarms (✓) None

## SCHEDULE OF ITEMS INSPECTED †*See note below*

### Protective measures against electric shock

**Basic and fault protection**

**Extra low voltage**  N/A SELV

**Double or reinforced insulation**  N/A Double or reinforced insulation

**Basic protection**
- ✓ Insulation of live parts
- ✓ Barriers or enclosures

**Fault protection**

**Automatic disconnection of supply**
- ✓ Presence of earthing conductor
- ✓ Presence of circuit protective conductors
- ✓ Presence of main protective bonding conductors
- ✓ Choice and setting of protective devices (for fault protection and/or overcurrent)

**Electrical separation**
- For one item of current-using equipment

### Additional protection
- ✓ Presence of residual current device(s)
- ✓ Presence of supplementary bonding conductors

**Prevention of mutual detrimental influence**
- ✓ Proximity of non-electrical services and other influences
- N/A Segregation of Band I and Band II circuits or Band II insulation used
- N/A Segregation of safety circuits

**Identification**
- ✓ Presence of diagrams, instructions, circuit charts and similar information
- ✓ Presence of danger notices
- ✓ Presence of other warning notices, including presence of mixed wiring colours
- ✓ Labelling of protective devices, switches and terminals
- ✓ Identification of conductors

**Cables and conductors**
- ✓ Selection of conductors for current carrying capacity and voltage drop
- ✓ Erection methods

### Cables and conductors (cont)
- ✓ Routing of cables in prescribed zones
- ✓ Cables incorporating earthed armour or sheath or run in an earthed wiring system, or otherwise protected against nails, screws and the like
- N/A Additional protection by 30mA RCD (where required, in premises not under the supervision of skilled or instructed persons)
- ✓ Connection of conductors
- ✓ Presence of fire barriers, suitable seals and protection against thermal effects

**General**
- ✓ Presence and correct location of appropriate devices for isolation and switching
- ✓ Adequacy of access to switchgear and other equipment
- ✓ Particular protective measures for special installations and locations
- ✓ Connection of single-pole devices for protection or switching in line conductors only
- ✓ Correct connection of accessories and equipment
- ✓ Selection of equipment and protective measures appropriate to external influences
- ✓ Selection of appropriate functional switching devices

## SCHEDULE OF ITEMS TESTED
- ✓ External earth fault loop impedance, $Z_e$
- N/A Installation earth electrode resistance, $R_A$
- N/A Continuity of protective conductors
- N/A Continuity of ring final circuit conductors
- ✓ Insulation resistance between live conductors
- ✓ Insulation resistance between live conductors and earth
- ✓ Polarity
- ✓ Earth fault loop impedance, $Z_s$
- N/A Verification of phase sequence
- ✓ Operation of residual current device(s)
- ✓ Functional testing of assemblies
- ✓ Verification of voltage drop

† *See note below*

Measured $Z_e$ 0.21 Ω

† *All boxes must be completed.* '✓' indicates that an inspection or a test was carried out and that the result was **satisfactory**. 'N/A' indicates that an inspection or test was **not applicable** to the particular installation.

‡ Where a smoke alarm has been installed, separate certification is required on the appropriate form.

This form is based on the model Electrical Installation Certificate shown in Appendix 6 of BS 7671 (as amended).
Published by NICEIC Group Limited © Copyright The Electrical Safety Council (Jan 2008)

**DOMESTIC INSTALLER**

# DOMESTIC ELECTRICAL INSTALLATION CERTIFICATE

DCP4/

**Original** (To the person ordering the work)

This certificate is not valid if the serial number has been defaced or altered

DOMESTIC INSTALLER

## CIRCUIT DETAILS

* To be completed only where this consumer unit is remote from the origin of the installation.

Record details of the circuit supplying this consumer unit in the bold box.

| Circuit number | Circuit designation | D = Distribution circuit F = Final circuit | Type of wiring (see code) | Reference method (see Appendix 4 of BS 7671) | Number of points served | Circuit conductors csa Live (mm²) | Circuit conductors csa cpc (mm²) | Max. disconnection time permitted by BS 7671 s | Overcurrent protective devices BS (EN) | Type No | Rating A | Capacity kA | RCD Operating current IΔn mA | Maximum Zs permitted by BS 7671 Ω |
|---|---|---|---|---|---|---|---|---|---|---|---|---|---|---|
| 1 | COOKER | F | A | 100 | 1 | 6 | 2.5 | 0.4 | 60009 | B | 32 | 6 | 30 | 0.44 |

## TEST RESULTS

| Circuit impedances (Ω) Ring final circuits only (measured end to end) r1 Line | rn Neutral | r2 cpc | All circuits (At least one column to be completed) R1 + R2 | R2 | Insulation resistance Line/Line MΩ | Line/Neutral MΩ | Line/Earth MΩ | Neutral/Earth MΩ | Polarity | Maximum measured earth fault loop impedance Zs Ω | RCD operating times at IΔn ms | at 5IΔn (if applicable) ms |
|---|---|---|---|---|---|---|---|---|---|---|---|---|
| | | | 0.33 | | | >200 | >200 | >200 | ✓ | 0.48 | 43 | 19 |

Location of consumer unit(s) **GARAGE**

Designation of consumer unit(s)

Prospective fault current at consumer unit(s) **0.8** kA

## TEST INSTRUMENTS

*Test instruments (serial numbers) used*

Multi-functional **N/A**

Insulation resistance **08H46**

Continuity **08H46**

Earth electrode resistance **N/A**

Earth fault loop impedance **076H90** RCD **740026**

This form is based on the model Electrical Installation Certificate shown in Appendix 6 of BS 7671 (as amended). Published by NICEIC Group Limited © Copyright The Electrical Safety Council (Jan 2008).

DCP4/5   Page 3 of 3

### CODES FOR TYPE OF WIRING

| | |
|---|---|
| A | PVC/PVC cables |
| B | PVC cables in metallic conduit |
| C | PVC cables in non-metallic conduit |
| D | PVC cables in metallic trunking |
| E | PVC cables in non-metallic trunking |
| F | PVC/SWA cables |
| G | XLPE/SWA cables |
| H | Mineral-insulated cables |
| O | Other - please state |

**DOMESTIC INSTALLER**

This safety certificate is an important and valuable document which should be retained for future reference

This certificate is not valid if the serial number has been defaced or altered

**DMP3/**

# MINOR DOMESTIC ELECTRICAL INSTALLATION WORKS CERTIFICATE

Issued in accordance with *British Standard 7671 – Requirements for Electrical Installations* by an Approved Contractor or Conforming Body enrolled with NICEIC, Warwick House, Houghton Hall Park, Houghton Regis, Dunstable, LU5 5ZX.

### To be used only for minor electrical work which does not include the provision of a new circuit

## PART 1: DETAILS OF THE MINOR WORKS

Client **MRS F G GRANT**

Date minor works completed **01-07-08**    Contract reference, if any **NONE**

Description of the minor works
**REPLACE EXISTING COOKER UNIT FOR UNIT WITH A 13A SOCKET OUTLET**

Details of departures, if any, from BS 7671 (as amended)
**NONE**

Location/address of the minor works
**3 BISHOPS CLOSE BATH SOMERSET**   Postcode **S036HT**

## PART 2: DETAILS OF THE MODIFIED CIRCUIT

| | | |
|---|---|---|
| System type and earthing arrangements | TN-C-S | TN-S ✓   TT |
| Protective measures against electric shock | ADS ✓ | Other |

Overcurrent protective device for the modified circuit   BS(EN) **61009**   Type **B**   Rating **32** A

Residual current device (if applicable)   BS(EN) **61009**   Type **B**   I∆n **30** mA

Details of wiring system used to modify the circuit   Type **PVC/PVC**   Reference method **100**   csa of live conductors **6** mm²   csa of cpc **2·5** mm²

Where the protective measure against electric shock is ADS, insert maximum disconnection time permitted by BS 7671 **0·4** s   Maximum $Z_s$ permitted by BS 7671 **1·44** Ω

Comments, if any, on existing installation   **NONE**

## PART 3: INSPECTION AND TESTING OF THE MODIFIED CIRCUIT AND RELATED PARTS   † Essential inspections and tests

| | |
|---|---|
| † Confirmation that necessary inspections have been undertaken | (✓) |
| † Circuit resistance   $R_1 + R_2$ **0·33** Ω   or   $R_2$ ___ Ω | |
| Insulation resistance (* In a multi-phase circuit, record the lower or lowest value, as appropriate)   Line/Line* ___ MΩ | |
| Line/Neutral **>200** MΩ | |
| Instrument Serial No(s)   † Line/Earth* **>200** MΩ | |
| † Neutral/Earth **>200** MΩ | |

| | | |
|---|---|---|
| † Confirmation of the adequacy of earthing | ✓ | (✓) |
| † Confirmation of the adequacy of protective bonding | ✓ | (✓) |
| † Confirmation of correct polarity | ✓ | (✓) |
| † Maximum measured earth fault loop impedance, $Z_s$ | **0·48** | Ω |
| † RCD operating time at I∆n (if RCD fitted) | **43** | ms |
| RCD operating time at 5 x I∆n, if applicable | **19** | ms |

Agreed limitations, if any, on the inspection and testing   **COOKER CIRCUIT ONLY**

## PART 4: DECLARATION

I/We CERTIFY that the said works do not impair the safety of the existing installation, that the said works have been designed, constructed, inspected and tested in accordance with BS 7671: **2008** (IEE Wiring Regulations), amended to ___ and that the said works, to the best of my/our knowledge and belief, at the time of my/our inspection complied with BS 7671 except as detailed in Part 1.

The results of the inspection and testing reviewed by the Qualified Supervisor

Name (CAPITALS) **R ROGERS**
Signature **R Rogers**
Date **1-07-08**

Name (CAPITALS) **T RIGGER**
Signature **T Rigger**
Date **1·7·08**

For and on behalf of *(Trading Title of Domestic Installer)*
**R AND T ELECTRICAL**
Address **6 COWBOY CLOSE BATH SOMERSET**   Postcode **S036QT**

Registration Number **1 0 9 8 7 6 5 4 3**   (The registration number is essential information)

Please see the 'Notes for Recipients' on the reverse of this page.

DMP3/1

SPECIMEN

# Appendix E

# Answers to questions

## TEST 1

1. Any values which are indicated as greater than can be disregarded as the true value is unknown.

$$\frac{1}{5.6} + \frac{1}{8.7} + \frac{1}{12} + \frac{1}{7} = 0.51 \qquad \frac{1}{0.51} = 1.92 \, M\Omega$$

Enter into calculator as:

$$5.6 \, x^{-1} + 8.7 x^{-1} + 12 x^{-1} + 7 x^{-1} = x^{-1} = 1.92$$

This is acceptable as the total insulation resistance is greater than $0.5 \, M\Omega$ and each circuit is greater than $2 \, M\Omega$.

2. (i) A $2.5mm^2/1.5mm^2$ cable has a resistance of $19.51m\Omega$ per metre. The resistance of 54 metres is:

$$\frac{54 \times 19.51}{1000} = 1.05 \, \Omega \qquad \frac{1.05}{4} 0.26 \, \Omega$$

$R_1 + R_2$ for the circuit is $0.26 \, \Omega$

$$Z_S + Z_e + R_1 + R_2 \qquad 0.24 + 0.26 = 0.5 \, \Omega$$

(ii) From Table 41B2 in BS 7671, the Maximum value for a type C32 amp device is 0.75. Use the rule-of-thumb to compensate for conductor operating temperature and ambient temperature.

$$0.75 \times 0.75 = 0.56 \, \Omega$$

Therefore this circuit will comply.

3. (i) From Table 9A in the *On-Site Guide*, the $r_1 + r_2$ value for the $4\,mm^2/1.5\,mm^2$ conductors is $16.71\,m\Omega$ /M. The $R_1 + R_2$ value at the sockets is:

$$\text{Socket 1:} \quad \frac{12 \times 16.71}{1000} = 0.2\,\Omega$$

$$\text{Socket 2:} \quad \frac{18 \times 16.71}{1000} = 0.3\,\Omega$$

$$\text{Socket 3:} \quad \frac{20.5 \times 16.71}{1000} = 0.34\,\Omega$$

$Z_S$ will be $0.34 + 0.6 = 0.94\,\Omega$. The maximum $Z_S$ for a 30A BS 3036 fuse from Table 41.2 in BS 7671 (0.4s) is given as $1.09\,\Omega$.

(ii) Using the rule-of-thumb for temperature correction the maximum permissible value is: $1.09 \times 0.8 = 0.87\,\Omega$. This value is acceptable.

4. (i) From Table 9A in the *On-Site Guide*, the $r_1 + r_2$ value for $10\,mm^2/4\,mm^2$ copper is $6.44\,m\Omega/M$.

$$R_1 + R_2: \quad \frac{6.44 \times 14.75}{1000} = 0.095\,\Omega$$

$Z_S$ for the circuit is $0.095 + 0.7 = 0.795$ (0.8) $\Omega$.

(ii) The maximum $Z_S$ for a 45A BS 3036 fuse from Table 4.4 in BS 7671 (5s) is $1.59\,\Omega$.

Corrected for temperature: $1.59 \times 0.8 = 1.27\,\Omega$. As the actual $Z_S$ is lower the circuit will comply.

5. (i) From Table 9A in the *On-Site Guide* $4\,mm^2$ copper conductors: have a resistance of $4.61\,m\Omega/M$.
$R_1 + R_2$ for the phase and CPC both in $4\,mm^2$ is $9.22\,m\Omega/M$

$$R_1 + R_2: \quad \frac{9.22 \times 67}{1000} = 0.061\,\Omega$$

As it is a ring:

$$\frac{0.61}{4} = 0.15\,\Omega$$

The $R_1 + R_2$ value is $0.15\,\Omega$.

$$Z_S = Z_e + R_1 + R_2$$

$0.63 + 0.15 = 0.78\,\Omega$ as the maximum permissible is given as $0.86\,\Omega$ this value is acceptable.

(ii)   As the maximum permissible $Z_S$ is given as $0.9$ and is taken from the *On-Site Guide*, no correction for temperature is required. We must subtract the $Z_e$ from the actual $Z_S$ to find the maximum permissible $R_1 + R_2$ value.

$$R_1 + R_2 = 0.9 - 0.63 = 0.27\,\Omega$$

We must now subtract the actual $R_1 + R_2$ value from the maximum permissible value.

$$0.27 - 0.15 = 0.12\,\Omega$$

The maximum resistance that our spur could have is $0.12\,\Omega$. To calculate the length we must transpose the calculation:

$$\frac{mV \times length}{1000} = R$$

to find the total length transpose to

$$\frac{R \times 1000}{mV} = length$$

Therefore:

$$length = \frac{0.12 \times 1000}{9.22} = 13\ metres$$

Total length of cable for the spur will be 8.67 metres.

6.   Electrical installation certificate
     Schedule of test results
     Schedule of inspection.

7. BS 7671 Wiring regulations
   *On-Site Guide*
   Guidance note 3
   GS 38.

8. Due date
   Clients request
   Change of use
   Change of ownership
   Before alterations are carried out
   After damage such as fire or overloading
   Insurance purposes.

9. Where there is recorded regular maintenance.

10.  (i)  P-N are $0.3\,\Omega$ each. The total loop will be 0.6 and the P-N at each
         socket after interconnection will be:

$$\frac{0.6}{4} = 0.15\,\text{ohms}$$

   (ii)  CPC must be $0.3 \times 1.67 = 0.5\,\Omega$
   (iii) $R_1 + R_2$ loop will be $0.5 + 0.3 = 0.8\,\Omega$. after interconnection the
         $R_1 + R_2$ value at each socket on the ring will be:

$$\frac{0.6}{4} = 0.2$$

11. $2.5\text{mm}^2/1.5\text{mm}^2$ has a resistance of $19.51\,\text{m}\Omega$ per metre. 5.8 metres of
    the cable will have a resistance of:

$$\frac{5.8 \times 19.51}{1000} = 0.113$$

    0.113 is the resistance of the additional cable. $R_1 + R_2$ for this circuit will
    now be $0.2 + 0.113 = 0.313\,\Omega$.

12. A statutory document is a legal requirement.

13. A non-statutory document is a recommendation.

14. Safety as the satisfactory dead tests ensure that the installation is safe to energize.

15. There are nine special locations.

16. The insulation resistance would decrease.

17.

$$\frac{1}{50} + \frac{1}{80} + \frac{1}{60} + \frac{1}{50} = 0.069 \qquad R = \frac{1}{0.069} = 14.45\,\Omega$$

18. Continuity of bonding and CPCs
Ring final circuit test
Insulation resistance test
Polarity
Live polarity at supply
Earth electrode ($Z_e$)
Prospective short circuit current
Residual current device
Functional tests.

19. Low resistance ohm meter
Low resistance ohm meter
Insulation resistance tester
Low resistance ohm meter
Approved voltage indicator
Earth loop impedance meter
Prospective short circuit current tester
RCD tester.

20. 15 mA, 30 mA, 150 mA (*only 150 mA if used for supplementary protection*).

21. Five times.

22. 0.05 Ω.

23. From Table 9A in the *On-Site Guide*, 10 mm² copper has a resistance of 1.83 mΩ per metre:

$$\frac{1.83 \times 22}{1000} = 0.4\,\Omega$$

24. TT system.

25. Socket 1: Good circuit
    Socket 2: CPC and N reversed polarity
    Socket 3: P and N reversed polarity
    Socket 4: P and CPC reversed polarity
    Socket 5: Spur
    Socket 6: Loose connection of N
    Socket 7: Good circuit.

26. From Table 41.2 $Z_S$ for a 5A BS 3036 fuse is 9.58 $\Omega$.
    The $R_1 + R_2$ value for 1 mm$^2$ copper from Table 9A is 36.2 m$\Omega$.
    Maximum resistance permissible for the cable

$$18.5 - 0.45 = 18.05\,\Omega$$

Maximum length of circuit is:

$$\frac{18.05 \times 1000}{36.2} = 498 \text{ M}$$

(*Problem with volt drop if the circuit was this long*).

27. TT: 21 $\Omega$
    TNS: 0.8 $\Omega$
    TNCS: 0.35 $\Omega$

28. Twelve socket outlets, one for each socket on the ring.

29. Unlimited number.

30. GS 38.

31. Taps, Radiators, Steel bath. Water and gas pipes, etc.

32. Steel conduit and trunking. Metal switch plates and sockets. Motor case, etc.

33. 4 mm$^2$.

34. 1 M$\Omega$.

35. 500 volts, 1 mA.

36. In possession of technical knowledge or experience or suitably supervised.

37. 500 Volts DC.

38. Flexible. Long enough but not so long that they would be clumsy. Insulated. Identified. Suitable for the current.

39. Finger guards. Fused. Maximum 4 mm exposed tips. Identified.

40. Approved voltage indicator or test lamp.

41. Fixed equipment complies with British standards, all parts correctly selected and erected, not damaged.

42. To ensure no danger to persons and livestock and that no damage occurs to property. To compare the results with the design criteria. Take a view on the condition of the installation and advise on any remedial works required. In the event of a dangerous situation, to make an immediate recommendation to the client to isolate the defective part.

43. Ensure the fault is repaired and retest any parts of the installation which test results may have been affected by the fault.

44. To ensure that all single pole switches are in the phase conductor. Protective devices are in the phase conductor. ES lampholders are correctly connected. The correct connection of equipment.

45. E14 and E27 as they are all insulated.

46. The top surface must comply with IP4X. The sides and Front IP2X or IPXXB.

47. By calculation $Z_S = Z_e + R_1 + R_2$. Or use a low current test instrument.

48. 0.37 kW (Regulation 552-01-01).

49. (a) Type C
    (b) Type D
    (c) Type C

50. Safety electrical connection do not remove. Voltage in excess of 230 volts where not expected. Notice for RCD testing. Where isolation is not possible by the use of a single device. Where different nominal voltages exist. Periodic test date. Warning non-standard colours.

51. $1.2 \times 0.75 = 0.9\,\Omega$
    $0.96 \times 0.75 = 0.72\,\Omega$
    $1.5 \times 0.75 = 1.125\,\Omega$
    $4 \times 0.75 = 3\,\Omega$
    $2.4 \times 0.75 = 1.8\,\Omega$

52. IP 4X

53. $10\text{mm}^2$.

54. Electrical installation certificate
    Schedule of test results
    Schedule of inspection

55. BS 7671 (Electrical Wiring Regulations)
    *On-Site Guide*
    GS 38
    Guidance note 3

56. Health and Safety at Work Act 1974
    Electricity Supply Regulations
    Electricity at Work Regulations 1989
    Construction Design and Management Regulations
    Building Regulation Part P
    (Appendix 2 of BS 7671 covers statutory regulations).

57. Brown (L1), Black (L2), Grey (L3) and Blue (N).

## Test 2

1.  (i) Electrical Installation Certificate.
    (ii) Schedule of Test Results.
    (iii) Schedule of Inspection.
    (iv) Periodic Inspection Report for the existing installation.

2. To monitor the condition of an electrical installation.

3. Electricity at Work Regulations 1989.

4. A person who possesses sufficient technical knowledge, relevant practical skills and experience for the nature of electrical work undertaken and is able at all times to prevent danger and, where appropriate, injury to him/herself and others.

5. For the safety of persons and livestock against the effects of electric shock and burns, to protect against damage to property by fire and heat arising from an installation defect. To confirm that the installation is not damaged or has deteriorated so as to impair safety and to identify any defects within the installation which may give rise to danger.

6. Use the correct equipment for the test being carried out.
   Ensure that it complies with GS 38.
   Have a full understanding of how to use the equipment.

7. (i) Insulation resistance test.
   (ii) Prospective fault current.
   (iii) RCD test.

8. (i) Earth fault loop impedance.
   (ii) RCD.
   (iii) Prospective short circuit current.
   (iv) Phase rotation.

9. Water, Gas, Oil line, etc.

10. Temperature.

11. Long lead method (method 2).

12. Trip hazard with a long lead.

13. 0.5 ohms.

14. (i) To check for any interconnections.
    (ii) To ensure that all conductors form a complete ring.
    (iii) To measure $R_1 + R_2$.
    (iv) To check correct polarity at each socket.
    (v) To identify spurs.

15. The conductors which are connected would not be the opposite ends of the ring.

16. Live conductors, and live conductors and earth.

17. That the installation is isolated and that any equipment vulnerable to damage or likely to give false readings is isolated or disconnected.

18. The rule of thumb.

19. To ensure that the circuit/installation has an earth, for safety reasons.

20. Prospective short circuit current is the maximum current that could flow between live conductors within an installation.

21. Prospective earth fault current is the maximum current that could flow between live conductors and earth within an installation.

22. The highest of the two values.

23. (i) When all circuits of the location meet the required disconnection times.
    (ii) All circuits within the location are additionally protected by 30 mA RCDs.
    (iii) All extraneous conductive parts within the location are effectively connected via the main protective equipotential bonding to the main earthing terminal.

24. (i) Earth rods or pipes.
    (ii) Earth tapes or wires.
    (iii) Earth plates.
    (iv) Underground structural metalwork embedded in foundations

25. Overload, Short circuit and earth fault.

Always try to answer the questions in full using the correct terminology; for example: if asked "which is the type of inspection to be carried out on a new installation", the answer must be: An initial verification. For the document required for moving a switch or adding a socket the answer must be: An electrical installation minor works certificate, not just minor works.

Do not waste time copying out the question and write as clearly as you can.

State three statutory documents relating the inspecting and testing of electrical installations.

- The Electricity at Work Regulations 1989
- The Health and Safety at Work Act 1974
- The Electricity Supply Regulations.

Remember: Always try to answer the questions in full using the correct terminology (for example: if asked "which is the type of inspection to be carried out on a new installation?" the answer must be: An initial verification.

For the document required for moving a switch or adding a socket the answer must be: An Electrical Installation Minor Works Certificate, not just minor works.

Do not waste time copying out the question and write as clearly as you can.

# Index